# Packet Switc
# and X.25 Net

## Simon Poulton

Pitman

PITMAN PUBLISHING
128 Long Acre, London WC2E 9AN

A Division of Longman Group UK Limited

© S. Poulton 1989

First published in Great Britain 1989

**British Library Cataloguing in Publication Data**

Poulton, Simon
    Packet Switching and x.25 networks
    1. Computer systems. Networks. Data
    transmission. Packet switching systems
    I. Title
    004,6'6

    ISBN 0-273-02986-X

**Printed in Great Britain at The Bath Press, Avon**

---

## Acknowledgements

There are a number of people who have provided invaluable assistance in
the preparation of this book, and to whom I wish to give my thanks.

Chief amongst these is Neil Matthew. Neil spent a great deal of time and
effort on a detailed reading of the manuscript, and found many places
where my brain had gone offline without telling the pen. His sheer hard
work is much appreciated. Neil also made numerous suggestions, both
stylistic and technical, which have made the book much better than it
otherwise would have been.

I am very grateful to my wife Judith, who spent many hours on the word
processor deciphering my handwriting and turning it into a presentable
document. She also suffered my preoccupation in this project without
complaint, and offered much needed support and encouragement.

Thanks are also due to Anselm Waterfield one of my colleagues, and to
Ian Campbell of Exeter University, who both read through the finished
work and gave me valuable reassurance.

Finally to Racal-Milgo Ltd., my employers, who loaned the equipment
for the photograph, and Camtec Electronics Ltd. who gave permission for
me to describe their Network Management System.

# Contents

# 1 The packet switching network

## 1.1 People and computers

When computers started to become business tools and to assume an essential role in all companies, they were very large both physically and in terms of capital and revenue cost. Indeed, when erecting a new building the company would have to design features especially for the computer. A big room to house it, separate power supplies, large air-conditioning plants, and hoisting gear to lift the bits of computer off a lorry into the computer room.

Once the computer was in place then it was treated very much like the directorial suite. It had its own retinue of staff who were somehow a different breed of people to the rest of the organization, and its own set of security procedures not only to prevent access but to deter enquiries.

The computer users – the accountants and engineers – would approach the building clutching a pile of punched cards or papertape and hand it over to an operator who would take them to the inner sanctum to be fed into the machine to be processed. Some hours later the results of this labour would by placed into a pigeonhole to be collected by the user.

In the early seventies Time Sharing started to become common. We now know this as accessing the computer via a terminal, though in those days it was more typically a mechanical teletype running at 50 or 110 baud, and even quite large installations would only be able to run half a dozen of them. Nevertheless this sent a shiver of excitement throughout the user community. It was possible for a department to have its own interface to this vast resource, and this confirmed a great degree of status. For the first time people felt computing power to be in their grasp and, although perhaps not driven by need, people wanted access and fought to get it.

By the mid-to-late seventies minicomputers were available and computing power started to become decentralized. The actual use of these machines, certainly at the start, was far less significant than their potential effect. For the best part of ten years there were bloody boardroom battles with Data Processing Managers (DPMs) fighting to save their traditional status and power, and everyone else was striving to make use of the machines and facilities that were becoming available. This is not to say that the DPMs were driven by anything other than

1

reasonable motives. A company with departmental minis and no centralized buying policy could easily find itself with a collection of different machines none of which were being run efficiently.

By the mid-eighties the users had won, mainly because of the technology revolution that had been going on largely in the background. The departmental mini then had as much power and access potential as the mainframe that the DPM was fighting to preserve.

As this battle raged a new issue came to the fore, that of networking. With the mainframe, as its number of terminals grew, there was an obvious and consistent network architecture, something like that in Fig. 1.1. The mainframe always has a star configuration. If terminals are added they are always a new ray from the centre whether they are connected directly, multiplexed, or via modems. The strategy is simple and easy to budget.

In the late seventies the same strategy applied to the minis in the organization. They were all treated as little mainframes and the network diagram changed to that shown in Fig. 1.2.

By the early eighties this was starting to creak a little. By this time users had expectations of technology and were starting to think for themselves about what they wanted to do. Users on their departmental minis could see advantages in being able to access the machine in

Fig. 1.1 Star networks based around mainframe computers

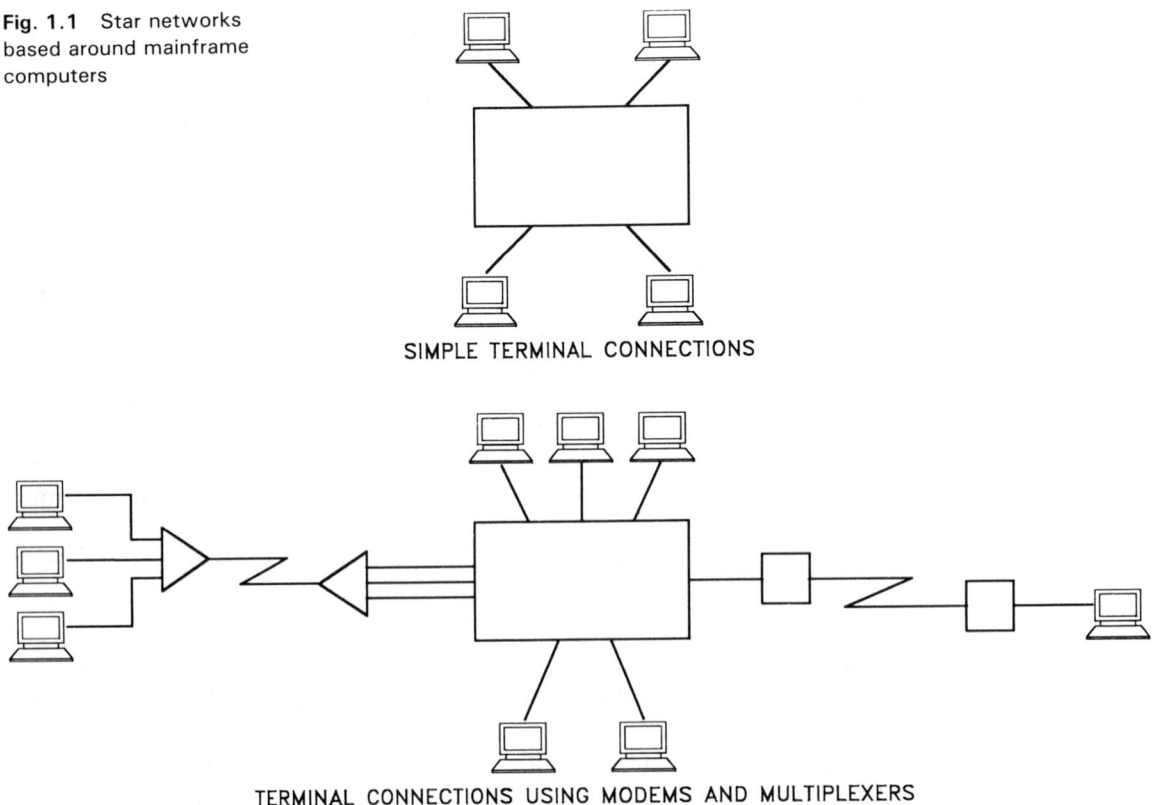

SIMPLE TERMINAL CONNECTIONS

TERMINAL CONNECTIONS USING MODEMS AND MULTIPLEXERS

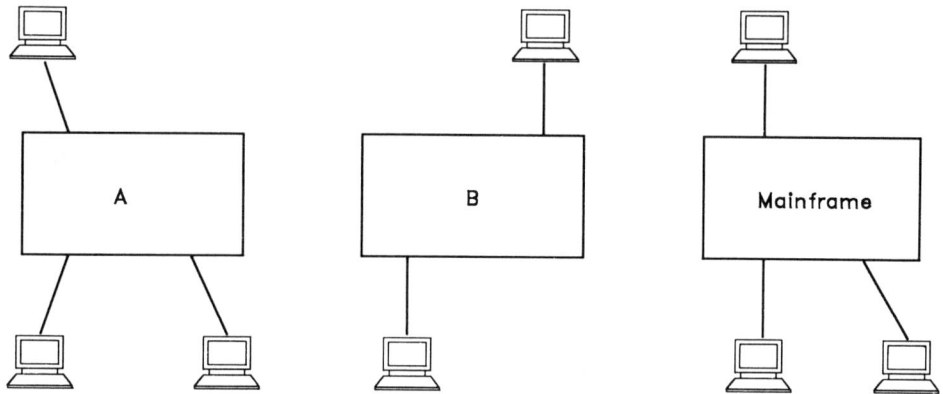

Fig. 1.2 Discrete
networks of mainframe
and minicomputers

another department as well. Perhaps because they had a software package to which access was occasionally required, or perhaps the user sometimes had a very large task which would swamp the mini and ought to run in the mainframe.

Suppliers of modems and multiplexers in star configurations were quick to spot this need and developed equipment to address it. The network then changed to that shown in Fig. 1.3. What this equipment did was to make the users connect to a Communications Processor rather than to a computer.

When beginning a session the user now got a message from the communications processor, and had to tell the processor which machine to connect to and what precise facilities were required. The processor had knowledge of all machines and passed a message along the trunk

Fig. 1.3 Connecting the
discrete networks together

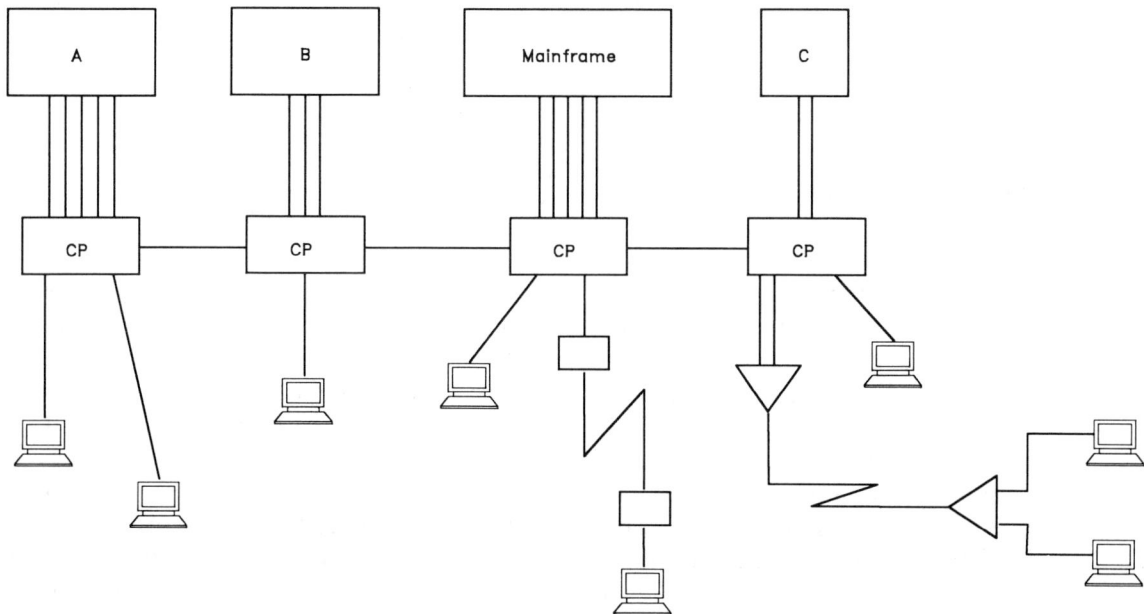

CP = Communications processor

line, perhaps through intervening processors to the destination processor. This then found a line to the required computer that conformed to the facilities profile that the user requested, and made a connection. Nowadays this concept is seen as natural, but it should be appreciated that this was all happening as the mini/mainframe battle was being fought, and represented a fairly major change in thinking.

Although this is somewhat less sophisticated than the systems around today – as will be shown – it has all the hallmarks of a Network. These are:

- All services (computers) and users (terminals) connect physically to the network, not to each other.
- The user has to tell the network which service is required to be accessed, and which facilities such as speed of access, amount of memory, and peripherals are required.
- The network determines whether the request can be met and, if so, allocates the service to the user.
- When the user has finished, the network has to reset everything that has been done back to its initial state.
- Both user and service are users of the network.

By the early eighties attention was being focussed on the network, as it was an area with high potential growth. It would also require international standardization to ensure that equipment from all manufacturers could be made part of a network.

Many types of architecture were developed to implement a network and many schemes of carrying data along wires were developed. One such architecture and scheme is Packet Switching which is the subject of this book. The book will explain and illustrate the subject with particular reference to X.25 networks.

## 1.2   Networking solutions

Before describing packet switching and networks that conform to international standards, it is instructive to take a brief look at other – older – technologies that have been used to connect users to the services that they want to access.

The use of multiplexers has already been mentioned as a tool that was in use right from the start of remote access to computers. The multiplexer is a straightforward data funnel that has several access points on one side, and an aggregate or trunk port on the other. The multiplexer then shares the transmission capability of the trunk between the terminals or host ports connected to it. The advantage of this scheme, as shown in Fig. 1.1, is that there is only a single communications link between the two multiplexers, and therefore there is only a single line rental to pay and a single pair of modems to buy. It is thus cost-effective to link say a branch office to a corporate headquarters, because several co-located users can all access the computer using a single link.

Multiplexers work on three basic principles: Frequency Division Multiplexing (FDM); Time Division Multiplexing (TDM); and Statistical Multiplexing (Stat. Mux.).

A frequency division multiplexer works by splitting the bandwidth of the cable into several narrow frequency ranges, and allocating a range to each communications channel. It is therefore very like the splitting of the airwaves into several radio channels. The technique is costly and is not suited to the limited bandwidth of telephone circuits, and is therefore very rarely used in data communications.

In time division multiplexing the device scans each of the channels in turn, and passes the selected data onto the aggregate. It is like a railway turntable which constantly takes a wagon from each of the branch rails and passes it to the main line. This can only work if the aggregate can handle the combined data of the channels; otherwise the multiplexer would have to stop the flow of data on the channels.

Most TDMs allow the aggregate bandwidth to be split unequally, so a 9600 bps link could support say two 2400 bps terminals and four 1200 bps printers. The two TDMs attached to a link must be configured so that each has the same understanding of how the time on the aggregate is allocated. This allows the individual channels to be derived correctly at the far end.

Statistical multiplexers work by having a buffer for each of the channels, and they allocate the aggregate according to how full the buffers are. Thus each terminal or host port "talks" to its own buffer, and does not know that the data is not being transmitted at that instant. The buffers are emptied into the aggregate at a speed relative to how full they are. It is like the postal service emptying letter boxes; each box is emptied regularly, but those that have a lot of letters are emptied more frequently.

The statistical multiplexer allows the sum of the individual channel speeds to exceed the aggregate link speed, because, when the multiplexer is not servicing a particular channel, the channel data is inserted into its buffer. To avoid the buffer filling, and the consequent necessity of the multiplexer stopping the channel, the average data rate of the channels should not exceed the data rate of the aggregate. This means that if a terminal runs at a speed of 9600 bps, but is only sending data one minute out of ten, then its effective data rate is 960 bps. Five such terminals could be statistically multiplexed down a 4800 bps aggregate.

Since there is no way of knowing which channel may be using the aggregate at any instant, it is necessary for the sending stat. mux. to precede each transmission with an indication of which channel the data is intended for. This allows the receiving stat. mux. to direct the data to the appropriate destination port. The indication forms an address which informs the destination how to route the data.

So far, all the devices mentioned connect two sites together, and data input to a given port on one site is always output from a known port at the other site. In the case of statistical multiplexers which must exchange addressing information anyway, these limitations can be removed.

If we consider Fig. 1.3, then the communications processors can be statistical multiplexers. The addressing information preceding data on the aggregates simply has to be extended to show the destination stat. mux. as well as the port. The intervening stat. muxes. can examine the destination field and route the data down the appropriate aggregate if it is not for them.

To allow the user to choose the destination, rather than to have it pre-configured by the Network Manager, it is only necessary for there to be a dialogue between the user and the stat. mux. The user simply indicates that all data should be preceded by the address for port X on stat. mux. Y, and the network of processors then routes all data accordingly.

Once this expansion had been made, then stat. mux. manufacturers quickly enhanced their machines to include a range of useful features:

- Destination Names so that the user can connect to FINANCE DATABASE and the processor looks up the actual network address in its configuration data.
- Hunt Groups so that, rather than trying to connect to a particular host port, the destination processor will simply allocate the next free port on the host.
- Mesh Networks so that there may be multiple aggregate links allowing the data load to be balanced, and offering resilience to faults.

These are in addition to Port Contention which is automatically provided by the statistical multiplexing concept. This provides the ability for there to be twenty users of ten host ports, although only ten can be working at once.

The major drawback of the statistical multiplexer is that different manufacturers have different techniques of performing it, so users are locked into a particular system. This also means that there can be no aggregate link into the host, only a multiplicity of low-speed connections.


## 1.3  Packet switching

In a Packet Switching Network, as the name implies, data is communicated between the user and the service in the form of packets. Before we can look at this type of network we therefore need to know that:

A **packet** is a structure containing some data to be communicated, an indication of where the data is going to (destination address), and an indication of where the data has come from (source address).

Having said that, it is clear that the function of the components in the network is to receive the packet, look at the destination address, and from its knowledge of the network send the packet to the next appropriate component. See Fig. 1.4.

**Fig. 1.4** A packet
switching network

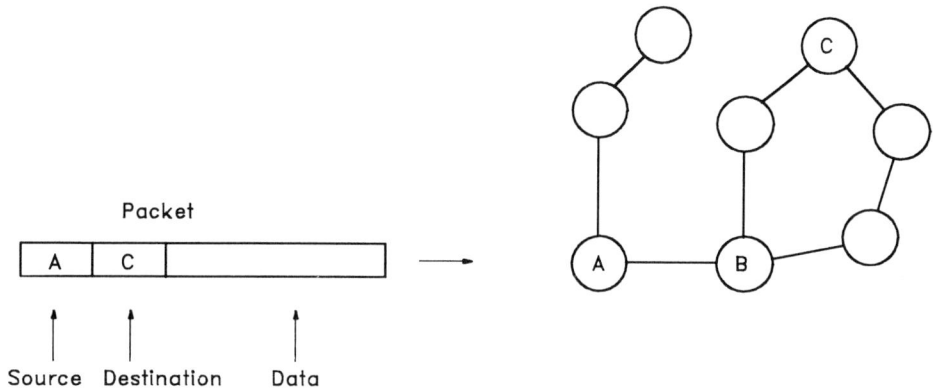

The packet has destination address C. Component A therefore must route the packet to component B. Component B has a choice; either of the onward routes is acceptable. The upper route looks preferable because there are fewer components; therefore the packet will suffer less delay. However, if the upper route is already being heavily used, or if it has a noisy, error-prone component, then the lower route is preferable. Whichever is chosen and for whatever reason the packet will arrive at component C as requested. When it does arrive, the receiver can extract the data and can determine the sender.

It is important to note that component C is a part of the network. It may not be a user or a service. For example if the packet is being routed to a screen, then only the data should appear, not the addresses. The network must deliver information appropriate to the end point, which is why both the service and the user of it must be considered users of the network.

When a user obtains access via the network to an accounting package in a computer, then neither the user nor the accounting package is concerned with addresses or the routing of packets. All of these are network functions, and both the user and the computer running the package are users of the network.

## 1.4 The layered network model

The above explanation of packet switching is simplistic because it does not address important issues such as the following:

● What is the format of the packet? How long are the fields in the packet and how are they encoded – Binary, ASCII, EBCDIC, etc?

7

- What is the style of addresses – numeric, alphabetic and so on. How are addresses assigned in the network, and how do you ensure there are no duplicates?
- What happens if the packet is corrupted in transit?
- What happens if the packet is completely lost in transit?
- What happens if there is more than one conversation going on?

Even if these issues are resolved in a particular network then how are the issues resolved in equipment from different manufacturers and how is connection between dissimilar networks achieved?

Fortunately, international standards bodies, such as the International Consultative Committee for Telephony and Telegraphy (CCITT), the Institute of Electrical and Electronic Engineers (IEEE), and the International Organization for Standardization (ISO) have addressed these issues and produced recommendations and standards. ISO have developed a model of a communication network which divides the problems into seven groups, and this is referred to as the Reference Model for Open Systems Interconnection. This is defined in the ISO 7498 standard. More concisely it is referred to as the ISO OSI Seven Layer Model.

The model is discussed in more detail in Chapter 6; however, for the moment we need only be concerned with the bottom three layers:

- *Layer 1*
  Covers the problems of getting a data bit from one component, via a transmission medium, to another component.
- *Layer 2*
  Covers the problems of ensuring that the bits are transferred with error notification.
- *Layer 3*
  Covers the problems of synchronizing the two users of the network.

The model only relates the functions of the layers and does not attempt to solve the problems. A network must therefore choose from a range of standards which are practical implementations of the narrow layers. Since the model defines what tasks are performed in each layer, any choice of standards is satisfactory as long as each standard conforms to the service required from the layer of the model.

In order that equipment from various manufacturers can interwork, and so that different networks can successfully be connected together, it is necessary that everyone chooses the same standards. There are therefore a limited number of network technologies, the most popular being X.25 and Ethernet. The international standards for these networks that conform to the OSI model will be discussed later.

In order to illustrate the functions of the layers we will examine X.25. Before doing so, a simple analogy. Consider a normal postal letter being carried in a postal van along the road between two postal depots. The letter is part of an individual conversation between the sender and the recipient, and we can imagine it as part of an ongoing series.

As part of the delivery process the letter has to be carried between various sorting offices. Each pair of offices will have individual control of the traffic flowing between them but will only be interested in the information written on the envelope, not the contents. The van on the road is only interested in the fact that it is going from A to B; it is not interested in the individual letters or their envelopes. In fact any carrier could be used equally effectively though the delivery times may be different.

This hierarchical arrangement – where each layer offers a service to the next – is similar to the way in which X.25 works.

### 1.4.1  Layer one

Layer one simply states how data bits are transferred. This could be implemented by emitting a squeak to indicate 0 and a squawk to indicate 1. Such a scheme would be less than satisfactory in most environments and the normal method is to send an electrical signal. Chapter 7 covers details of what happens, but essentially there are two signal paths between each two components. One signal path is for data in one direction, and the other path is for data in the other direction. The voltage level determines whether a 0 or 1 is indicated. Having two signal paths allows data to be sent in both directions simultaneously.

X.21 is the electrical system used and is a synchronous method of transmission. This is examined in Chapter 7, but for the moment this only needs to have one consequence to us which is that there is a continual "beat" called the *clock* and it is compulsory to send a bit at every beat.

### 1.4.2  Layer two

X.25 layer two defines methods for detecting and correcting errors in the transmission of bits from one component to another. In fact this is not precisely true as will be shown in Chapter 6, but it is nearly always the case.

At this point we have to look at a network with more than one conversation taking place (see Fig. 1.5). If there is a conversation between A and P, and one between B and Q, then it is clear that on the wire between M and N the two conversations have to be multiplexed. The multiplexing is nothing to do with layer two. Layer two is only concerned with the correct transmission of data along a wire; it is only concerned with the carrying of data between M and N. The combination of layer one and layer two provides a link between the nodes of the network, and layer two is often referred to as the *data link layer*.

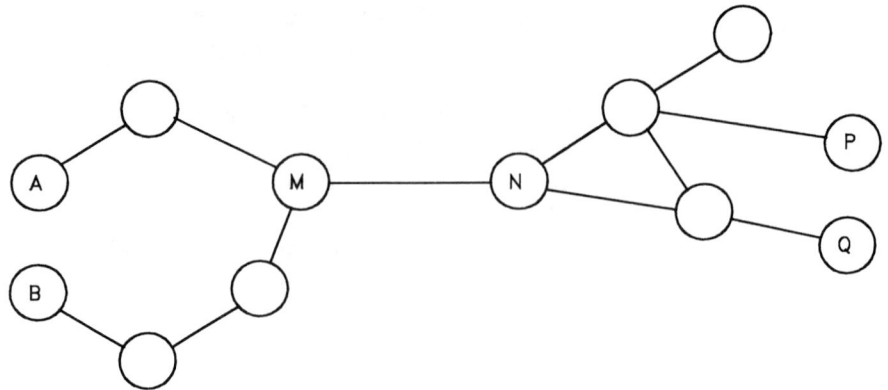

Layer two sends the data in a defined structure called the *frame*.

● The frame is the unit of transmission for layer two.
● The frame is contiguous – that is, once the sender starts to send a frame it will carry on to the end or give up completely and try again.

There is a further feature imposed by most hardware implementations that is not actually imposed by the standard:

● The frame is organised in bytes and always contains a whole number of bytes.

Since layer one requires that a bit is sent on every beat of the clock, layer two requires something to send between frames. This is the *flag byte* which is a set sequence of bits which the receiver must essentially ignore. Remembering that layer one is full duplex we can look at our example in more detail. See Fig. 1.6.

**Fig. 1.6** Flag bytes and frame delimiting

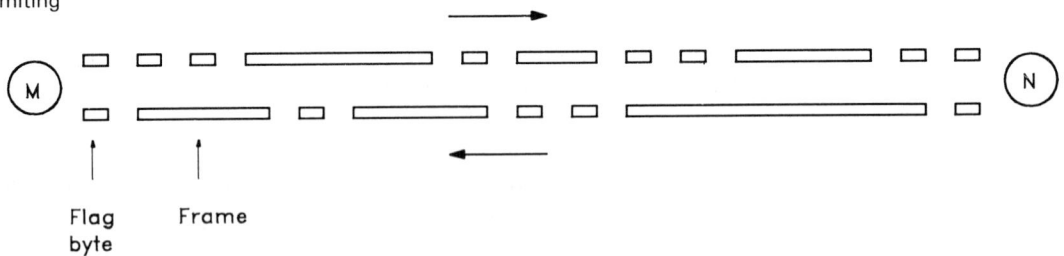

Flag    Frame
byte

Notice that there must always be a flag byte to delimit one frame from the next. It acts as a delimiter as well as an idle condition. Notice also that the frames are of arbitrary length. The layout of the frame is fixed and is as shown in Fig. 1.7.

The first byte of the frame is the *Address* byte, and simply indicates whether the frame is a command or response. This is explained in more detail in Chapter 4.

**Fig. 1.7** Format of the
layer two frame

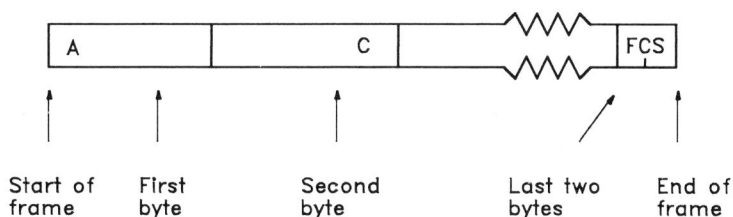

Start of        First        Second        Last two        End of
frame           byte         byte          bytes           frame

The final two bytes are called the *Frame Check Sequence* (FCS) and
are a value related to the contents of the frame. The field is like a
checksum, but is actually a Cyclic Redundancy Check. Using the FCS
the receiver of the frame can determine whether or not the frame has
been corrupted during transmission.

Since the frame is of variable length with no length indication, the
receiver must capture all of the data up to the terminating flag byte,
before it can determine the length of the frame and where the FCS is
located. It then calculates the FCS itself, and checks whether this agrees
with the FCS in the frame to determine whether corruption has
occurred.

The FCS does not catch all errors. Being 16 bits long it has 64 000
possible values. Thus if corruption occurs during transmission there is
some chance that it will not be detected if the FCS happens to be correct.

The second byte of the frame is the *Control* byte and indicates what
type of frame this is. The control byte therefore indicates the format of
the frame, different frames having different formats. One of the frame
types is the *Info* frame and is used to carry data from one node to
another. The layout of the Info frame is shown in Fig. 1.8.

**Fig. 1.8** Conceptual
layout of the Info frame

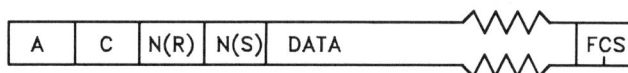

The third and fourth fields of the Info frame are the Send and Receive
sequence numbers which are explained below. They are followed by the
data. Note that there is no length indication. The length is determined
by the position of the terminating flag byte. The sequence numbers are
actually included in the Control byte, but in the early part of this book
they will be shown separately for clarity.

The *Send Sequence Number,* depicted N(S), indicates which number
this is in a series of Info frames. The first Info frame sent will have an
N(S) of 0, the next will have an N(S) of 1, and so on. This mechanism
allows the receiver to verify that it has received all of the frames and that
none have been lost in transit. The N(S) can rise to a value of seven and
will then cycle back to zero. This limits the length of the N(S) field to
three bits in the frame. There is a problem here in that if the N(S) can
cycle, and if eight frames are lost in transit and the next frame is sent
correctly, then the receiver cannot detect the loss. To prevent this, the

receiver has to acknowledge receipt of the Info frames, and the sender cannot send any new frames if seven are outstanding and awaiting acknowledgement.

The acknowledgement is carried in the $N(R)$ field of returning Info frames, and indicates the next frame that is expected.

Suppose a frame is sent with $N(R)$ of 3 and $N(S)$ of 7:

● The Receiver of the frame must check whether frame seven is the next one that it was expecting.
● The Receiver may now use an $N(S)$ of 3 again.

It is not necessary for every frame to be individually acknowledged. Fig. 1.9 shows an example conversation where only some of the frames are acknowledged, but the acknowledgement includes earlier frames as well.

Suppose that a user has requested that a file be listed out on screen and that the conversation is over a packet switched network. During the listing the data will be essentially unidirectional and the receiver in the

**Fig. 1.9** A layer two conversation; note that it is not necessary for every frame to be individually acknowledged

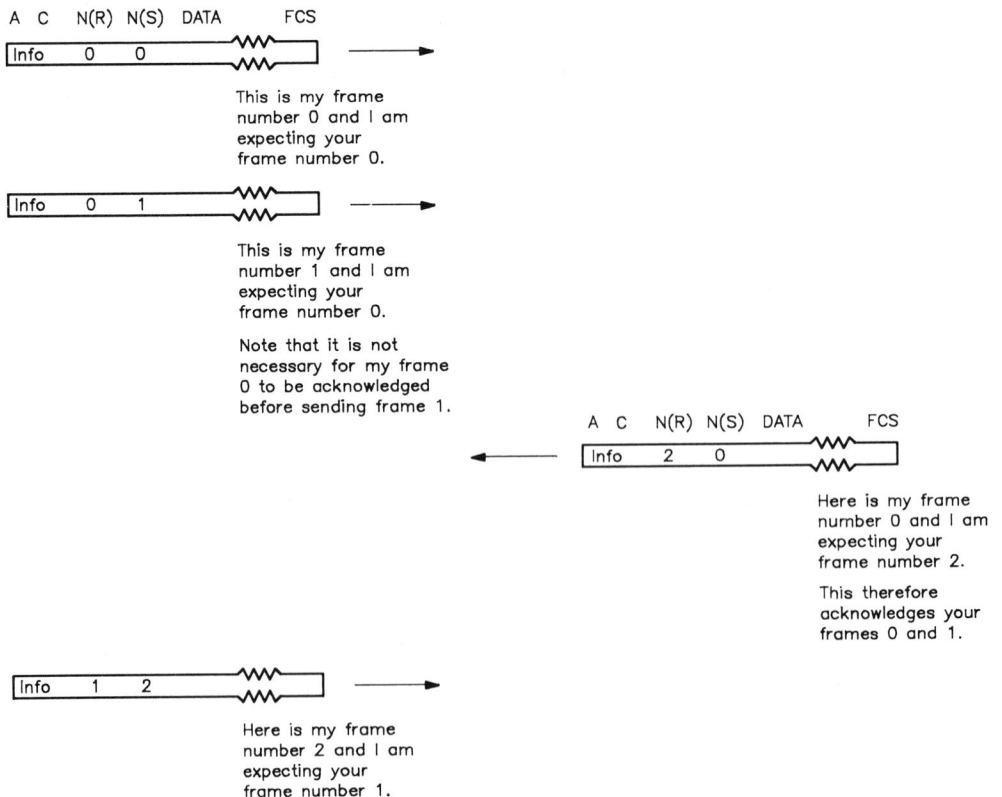

```
A  C   N(R) N(S)  DATA      FCS
  ┌────────────────────WWW──────┐
  │Info   0    0           WWW   │  ──────────▶
  └──────────────────────────────┘
```

This is my frame number 0 and I am expecting your frame number 0.

```
  ┌────────────────────WWW──────┐
  │Info   0    1           WWW   │  ──────────▶
  └──────────────────────────────┘
```

This is my frame number 1 and I am expecting your frame number 0.

Note that it is not necessary for my frame 0 to be acknowledged before sending frame 1.

```
                          A  C   N(R) N(S)  DATA      FCS
            ◀──────────   ┌────────────────────WWW──────┐
                          │Info   2    0           WWW   │
                          └──────────────────────────────┘
```

Here is my frame number 0 and I am expecting your frame number 2.

This therefore acknowledges your frames 0 and 1.

```
  ┌────────────────────WWW──────┐
  │Info   1    2           WWW   │  ──────────▶
  └──────────────────────────────┘
```

Here is my frame number 2 and I am expecting your frame number 1.

network needs to be able to acknowledge the receipt of Info frames without sending Info frames back. This is achieved with the *Receiver Ready* (RR) frame which just sends the N(R) value to the sender. Fig. 1.10 shows an example of this.

Unless the data is unidirectional, or unidirectional for a period, then there is no need for the use of the RR frame. However, most X.25 implementations require that frames are acknowledged within a short time of being sent, and the RR frame is very common.

The frame numbering scheme gets a little more complex when the numbers wrap around. The same logic still applies though. See Fig. 1.11.

We now need to look at what happens if there is a loss of data on the link. Consider the conversation shown in Fig. 1.12. Clearly something has gone awry. The most likely cause is that Info frame 4 from the lefthand side has been corrupted in transit, thus when it arrives at the

**Fig. 1.10** Use of the RR frame to acknowledge receipt of data without sending more data in return

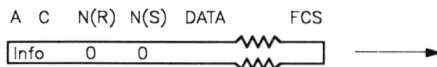

A   C   N(R)  N(S)  DATA        FCS

| Info | 0 | 0 |

Here is my frame 0.
I expect your frame 0.

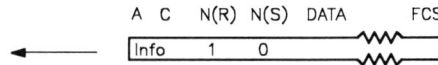

A   C   N(R)  N(S)  DATA        FCS

| Info | 1 | 0 |

Here is my frame 0.
I expect your frame 1.

This acknowledges
your frame 0.

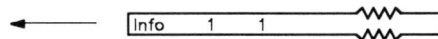

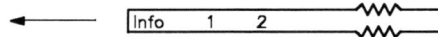

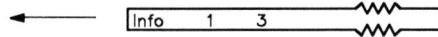

| Info | 1 | 1 |

| Info | 1 | 2 |

| Info | 1 | 3 |

Here are my frames
1, 2, and 3. I still
expect your frame 1.

| RR | 4 |

I am expecting your
frame number 4. This
therefore acknowledges
your frames 0, 1, 2, and 3.

A   C   N(R)  N(S)   DATA        FCS
Info      6     4

Here is my 4.
I expect your 6.

A   C   N(R)  N(S)   DATA        FCS

Info      5     6

Info      5     7

Here are my 6 and 7.
I expect your 5.

RR      0

I expect your 0.
This acknowledges
your 6 and 7.

Info      0     5

Here is my 5.
I still expect your 0.

Info      0     6

Info      0     7

Here are my 6 and 7.
I still expect your 0.

Info      0     0

Here is my 0.
I expect your 0.

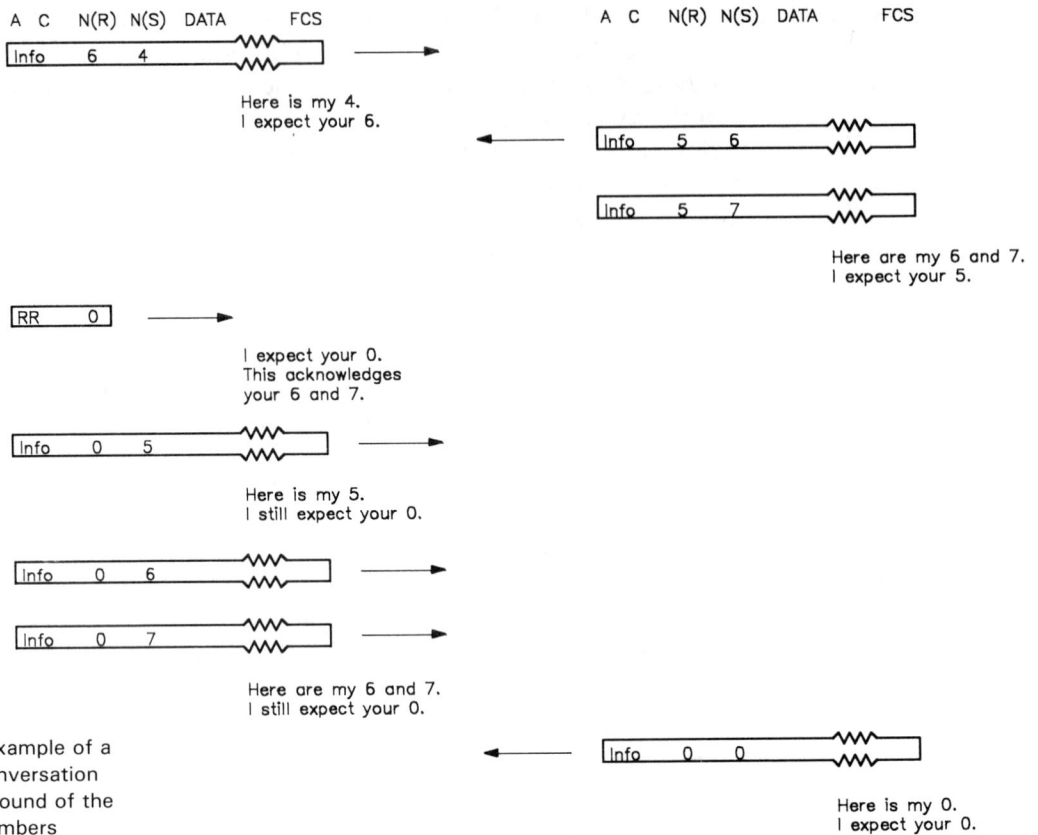

**Fig. 1.11**  Example of a layer two conversation with wrap-around of the sequence numbers

**Fig. 1.12**  Loss of frames in transit; note that it is unusual to lose the whole frame; normally it is corrupted, thus has a bad FCS and is therefore ignored by the hardware and not passed to layer two

receiver the FCS is incorrect and the whole frame is discarded without further attention. This is normally done by the hardware

What happens in these circumstances is that the righthand side will issue a *Reject* frame. This is done on receipt of an out-of-sequence frame – in this case number five – and will indicate where retransmission is to start. In this case we require frame 4 to be retransmitted. However, all subsequent frames will also be retransmitted as the receiver will ignore frames following the one that was out-of-sequence. The conversation would therefore continue as shown in Fig. 1.13.

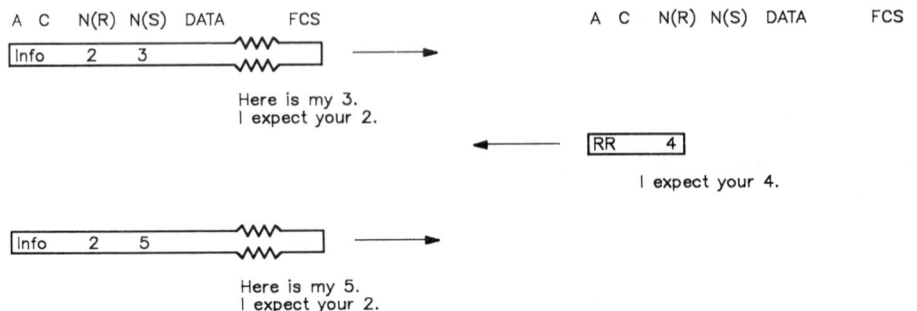

A   C   N(R)  N(S)   DATA        FCS
Info      2     3

Here is my 3.
I expect your 2.

A   C   N(R)  N(S)   DATA        FCS

RR      4

I expect your 4.

Info      2     5

Here is my 5.
I expect your 2.

14

A C   N(R) N(S)  DATA        FCS
| Info | 2 | 3 |

A C   N(R) N(S)  DATA        FCS

RR    4

Info | 2 | 5

REJ    4

Info | 2 | 4

Info | 2 | 5

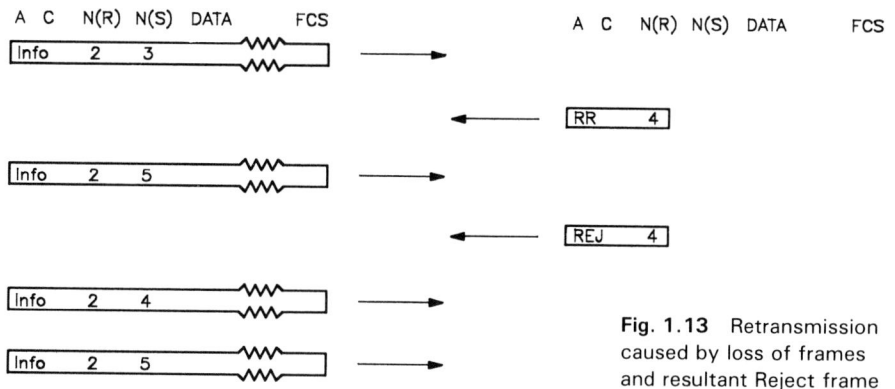

**Fig. 1.13** Retransmission caused by loss of frames and resultant Reject frame

Consider the conversation in Fig. 1.14. If we were now to wait for the time required by the network, then the righthand side would acknowledge receipt of all the data by sending an RR with N(R) of 4. Suppose that the timeout has not expired and the lefthand side has more data to send, what will happen? The lefthand side cannot send frame 4 because its frame 5 has not been acknowledged. If it were to send frame 4 and receive an RR with N(R) of 5 then either:

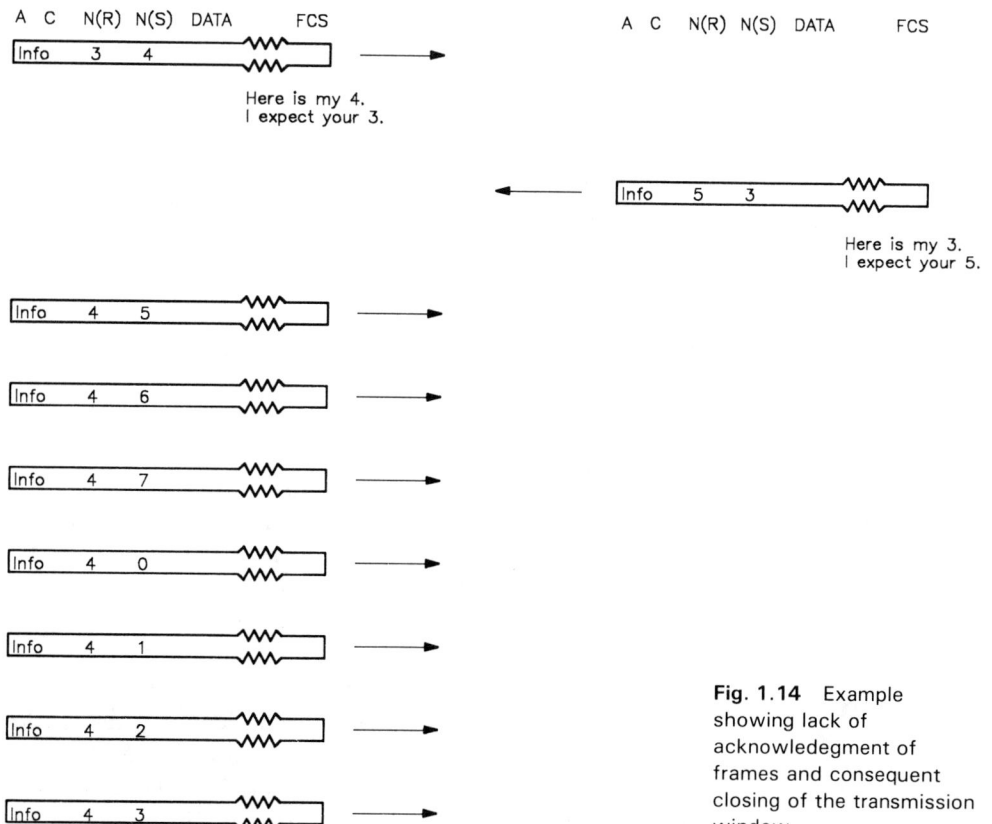

A C   N(R) N(S)  DATA     FCS
Info | 3 | 4

Here is my 4.
I expect your 3.

A C   N(R) N(S)  DATA     FCS

Info | 5 | 3

Here is my 3.
I expect your 5.

Info | 4 | 5

Info | 4 | 6

Info | 4 | 7

Info | 4 | 0

Info | 4 | 1

Info | 4 | 2

Info | 4 | 3

**Fig. 1.14** Example showing lack of acknowledegment of frames and consequent closing of the transmission window

- the receiver is acknowledging all eight frames, or
- the receiver has not received any of the eight frames and is just "reminding" the sender.

This confusion is obviously not acceptable so no more frames can be sent. The sender can send an RR — in this case with N(R) of 4 — to "remind" the receiver that some acknowledgement is required. Indeed many implementations regularly exchange RR frames to verify that both ends of the link are up and running.

The number of frames that can be outstanding awaiting acknowledgement is referred to as the *Window*, and this is normally seven as shown. Note that the window may be say set to three, but the N(R) and N(S) values still cycle with modulo eight. When all the frames allowed by the window are awaiting acknowledgement, then the window is said to be closed or full. When some frames can still be sent the window is said to be open

X.25 frames only have a three-bit field for N(R) and N(S) so windows of greater than seven are not possible. An extended format of the frame has a larger field and allows the values to cycle to 127 with a corresponding increase in possible window sizes. The extended format is of most use in satellite links where the time delays mean that many frames may be in transit at one time.

**The flag byte and bit stuffing**  No matter which of the possible 256 values is chosen for the flag byte, it is clear that this imposes a limitation on what can appear in the frame. If the flag value appears in the data portion of an Info frame then this will terminate the frame and it can never be transmitted. This problem is avoided by selecting the flag byte to be hex 7E, and by the process of bit stuffing. In binary the flag byte appears as

0111   1110

It therefore contains a string of six contiguous 1 bits. Bit stuffing takes the frame before transmission and, whenever there is a string of five bits, a 0 bit is "stuffed" in after them. There can therefore never be a string of six 1 bits in the frame, and thus there is no confusion over flag bytes and frame bytes. This is shown in Fig. 1.15.

Note that every sequence of five 1 bits is bit-stuffed whether or not it is followed by a 1, so that on reception every bit-stuffed zero is reliably removed. Note also that, though the frame is an exact number of bytes, after bit stuffing this may no longer hold true, so during transmission the frame is simply a string of bits.

At the receiver the reverse process to bit stuffing takes place. The flag bytes are removed from the bit stream leaving the "stuffed" frame. The bit string is then scanned and whenever five 1 bits are found the following 0 bit is stripped-off. The bit-stripping process results in the original frame — a whole number of bytes — being restored.

Fig. 1.15  Bit-stuffing

FRAME FOR TRANSMISSION — WHOLE NUMBER OF BYTES

| 0101 1001 | 0111 1110 | 0011 0111 | 1111 0000 | 0011 1110 |

Final bit to
be transmitted

First bit to
be transmitted

FRAME AFTER BIT—STUFFING — NO LONGER BYTE—ALIGNED

| 0101 1001 01011 1110 0011 01101 1111 0000 00011 1110 |

FRAME AS TRANSMITTED — BIT—STUFFED WITH FLAG BYTES

| 0111111001011001010111100011011011111000000011111001111110 |

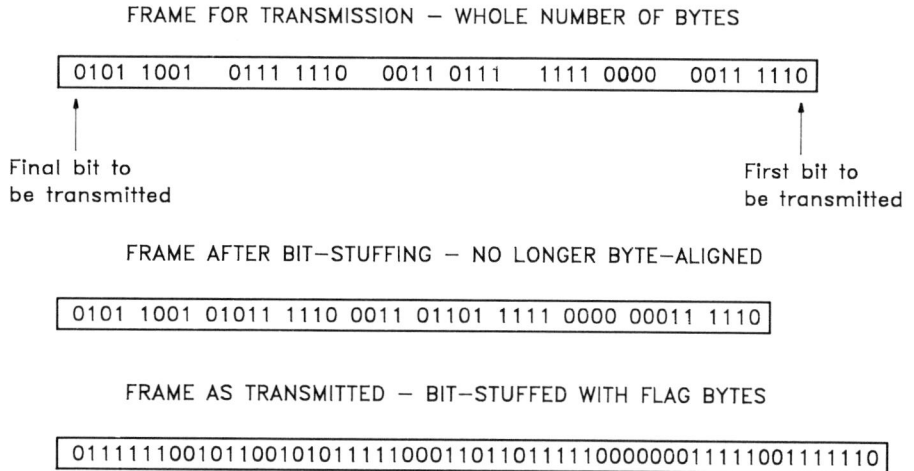

Layer two ensures that, between two nodes in the network, data can be transmitted and any errors are detected and corrected by retransmission. Layer two effectively gives an error-free network, but does not give any way of carrying several conversations at the same time. This is a function of layer three.

### 1.4.3  Layer three

Layer three — the *network* or *packet layer* — takes care of individual conversations in a network. It effectively takes care of what happens between any two end points in the network, whereas layer two takes care of what happens between any two nodes. See Fig. 1.16.

The simple network contains eight nodes to which four network users are connected. A conversation between points A and C will involve separate layer two links from A to P, P to Q, Q to R, R to S, and S to C. Each link will carry completely separate frames and each will maintain different sequence numbers. The links may run different window sizes. Even though each of the links ensures that data is carried with error detection and correction between the nodes, it does not ensure that the data is carried correctly within the network.

Fig. 1.16 Multiple conversations in the network; the conversations are between end points A and C, and B and D, and use the intervening layer two links

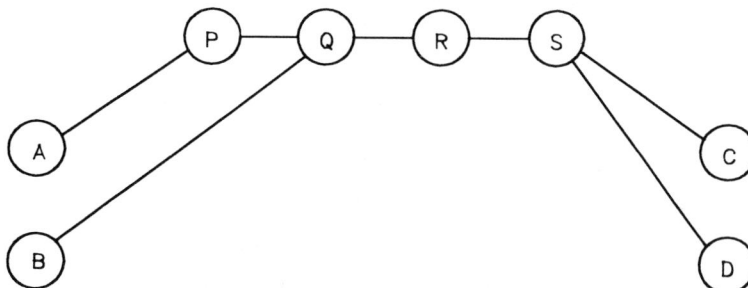

This can be appreciated by considering what happens if a wire breaks. Layer two will not be able to recover from the error—it cannot reestablish a broken wire. This fact cannot be reported onwards by the other layer two links in the circuit since each of them is still intact and functioning correctly.

The error has to be detected and corrected by some higher level of authority that sees the overall picture. To ensure that data is delivered across the network correctly, layer three takes care of the end-to-end connection and performs error detection and correction.

Going back to Fig. 1.16, layer three provides error detection and correction between points A and C and uses several layer two links. Suppose that there is a separate, simultaneous connection between points B and D; this will be subject to the same process. The network would then be carrying two layer three connections corresponding to the real conversations, and several layer two links. The Q to R and R to S links each have a simple layer two link which carries two layer three conections at the same time. The conversation may look something like that in Fig. 1.17.

**Fig. 1.17** Conceptual frame contents to allow multiple layer three conversations

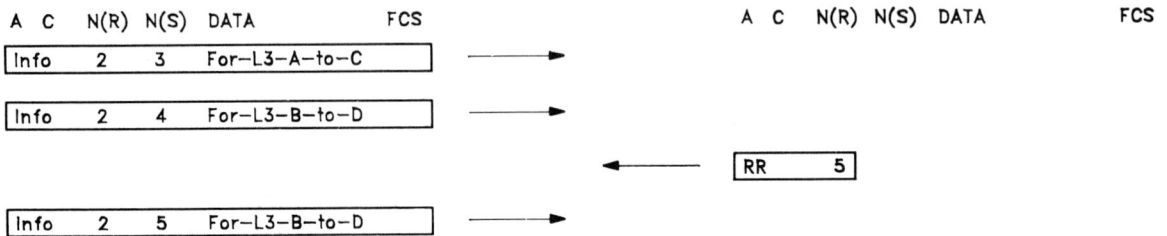

```
A  C   N(R) N(S)  DATA              FCS        A  C   N(R) N(S)  DATA              FCS
Info   2    3     For—L3—A—to—C          ───────▶
Info   2    4     For—L3—B—to—D          ───────▶
                                        ◀───────      RR        5
Info   2    5     For—L3—B—to—D          ───────▶
```

Many of the requirements of layer three can be inferred from what we have seen of layer two and from the need to carry several layer three conversations over a single layer two link.

- There has to be an identification of which layer three conversation the data in this frame refers to.
- There has to be some method of ensuring that all data arrives in sequence. This is done by P(R) and P(S) values which are akin to the N(R) and N(S) sequence numbers of layer two.
- There has to be a means of error recovery.

These items are carried in a data structure called the *packet* which is reminiscent of the layer two frame. As in layer two, there are several types of packet allowing the layer three conversation to proceed with all the required features. The packet type for carrying data is the *Data packet* as shown in Fig. 1.18. This is a conceptual illustration and does not show the actual packet layout.

There are two points to note:

- The LCI (*Logical Channel Identification*) field identifies which of the possible conversations this is.

Fig. 1.18 Conceptual layout of the Data packet

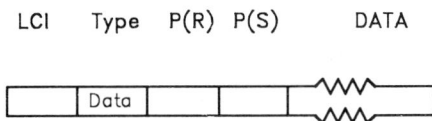

| LCI | Type | P(R) | P(S) | DATA |
| --- | --- | --- | --- | --- |

```
 _____
|      | Data |      |      |    ~~~~~~~  |
|_____|_____|_____|_____|____~~~~~~__|
```

- There is no length indication or termination of the data field. This is determined from the layer two frame format, or rather what is delivered by layer two.

Returning to Figures 1.16 and 1.17 we can now illustrate the conversation fully (see Fig. 1.19).

| Frame | A | C | N(R) | N(S) | DATA | | | | | FCS |
| --- | --- | --- | --- | --- | --- | --- | --- | --- | --- | --- |
| Packet | | | | | LCI | Type | P(R) | P(S) | DATA | |
| Info | | 2 | | 3 | 400 | Data | 0 | 0 | | |
| Info | | 2 | | 4 | 401 | Data | 0 | 0 | | |
| Info | | 2 | | 5 | 401 | Data | 0 | 1 | | |

Fig. 1.19 Frame layouts for the conversation showing the layer three Data packet within the layer two Info frame

The LCI field is expressed as three hex digits and is assigned arbitrarily by the two participating layer two nodes. In the example the first LCI to be used was 400 and the next 401. Each LCI is a separate conversation and has its own layer three P(R) and P(S) values. The two nodes will be configured with a range of LCIs that may be used.

The LCI is specific to each link, and several different LCIs may be used for the same conversation on the different links across the network. This is shown in Fig. 1.20.

If any of the conversations were to close down then the LCIs in use would be deallocated and would be available for new conversations.

The LCI is actually made up of two parts:

- The Logical Channel Group (LCG) which is the first hex digit.
- The Logical Channel Number (LCN) which is the last two digits.

It is normal to refer to the LCI as the LCN, even though this properly only identifies part of the data. Network administrators usually impose some structure on the LCN, and this is discussed in Chapter 4.

There are two ways in which LCNs may be allocated:

- *Permanent connections*
  In this method the LCNs are allocated automatically when the network components start up. The components have to be configured to make appropriate connections for the desired layer three conversation. This method would be used where a "permanent" conversation was required such as between two host computers that often exchange data.

● *Non-permanent connections*

In this method the LCNs are allocated dynamically as in Fig. 1.20 when the conversation starts. This method is suitable for terminal-to-host conversations where it is appropriate to allocate network resources only when they are required. The disadvantage of this approach is that there can be a delay whilst the connections are established, during which data cannot be transferred.

In general the non-permanent approach is much preferred, and where a permanent conversation is required then it is simply established once and left allocated. This is simply a decision of network suppliers who choose not to provide both methods.

To establish connections across a network that does not implement permanent connections, a *Call packet* is issued. This does several things:

● The Call packet contains the destination and source addresses, so can be routed by the nodes through the network.
● As the packet is routed throught the network the LCNs are allocated to the conversation which we can now give its proper name – the Call.
● When the packet arrives at the destination a *Call Accept packet* (CAA) is issued which returns through the network and indicates successful establishment of the call.
● Data can now be exchanged.

The connection is referred to as a *Virtual Circuit*. It is not an actual physical end-to-end circuit as would be the case with a telephone call. The permanent and non-permanent connections are referred to as Permanent Virtual Circuit (PVC) or Switched Virtual Circuit (SVC) respectively.

It is important to realize that the source and destination addresses in the Call packet are completely separate from the LCNs that may be allocated. It is only necessary that the nodes should be able to interpret them and know which physical link to route the call along. In the case of an X.25 network each address may be from 0 to 15 decimal digits.

When the conversation – or Call – is finished, for instance a user logs-off from a host, then either end can issue a *Clear Request packet* (CLR). This indicates to the other end that the call is finished. The receiver will "tidy-up" after the call and issue a *Clear Confirmation packet* (CLC). As the confirmation travels through the network, the network components release the resources allocated to the call and pass the confirmation on. Finally the CLC arrives at the end that sent the CLR and that end can tidy up after the call and all vestiges of it are then gone.

The layer three call, therefore, has three phases:

● Call establishment – the exchange of CALL and CAA.
● Data transfer.
● Call clear-down – the exchange of CLR and CLC.

**Fig. 1.20** Dynamic allocation of LCNs to conversations

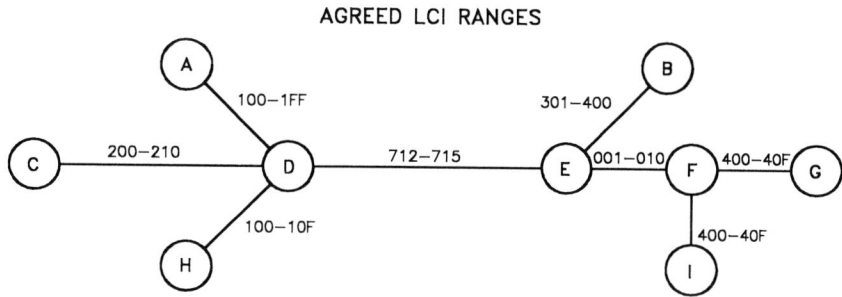

AGREED LCI RANGES

CONVERSATION BETWEEN A AND B — LCI'S ASSIGNED

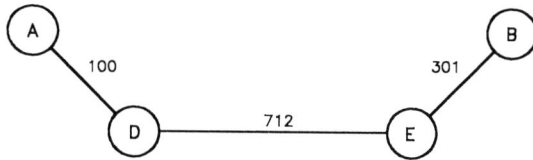

SECOND CONVERSATION ESTABLISHED BETWEEN H AND I

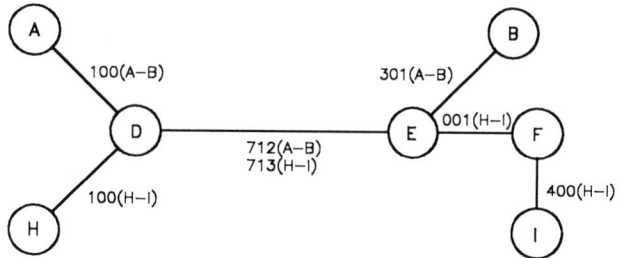

THIRD CONVERSATION ESTABLISHED BETWEEN C AND G

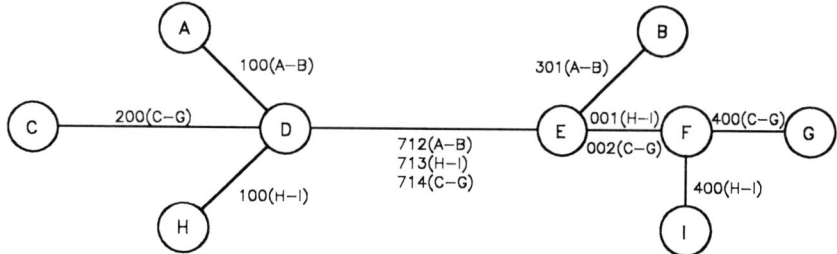

**Fig. 1.21** A short call showing call establishment, data transfer, and call clearing

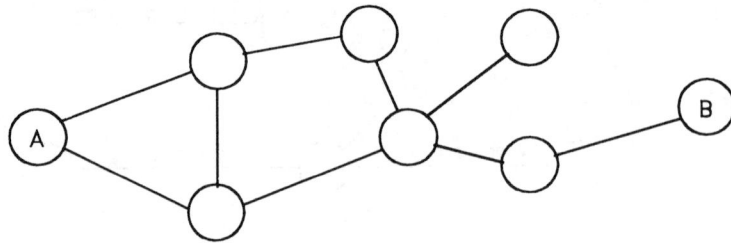

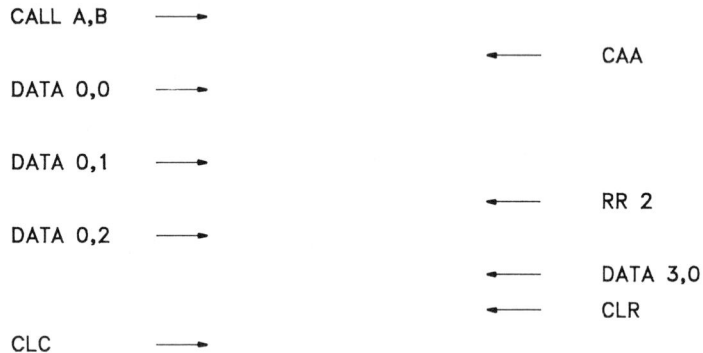

```
CALL A,B      ────▶
                           ◀────      CAA
DATA 0,0      ────▶
DATA 0,1      ────▶
                           ◀────      RR 2
DATA 0,2      ────▶
                           ◀────      DATA 3,0
                           ◀────      CLR
CLC           ────▶
```

Fig. 1.21 shows a short call.

If we examine the call setup phase at any of the nodes then we see a situation like that in Fig. 1.22. Suppose that at node N we already have calls in progress between P and Q. The Call packet arrives at the node from A and is to be routed to B. The internal configuration will be something like that shown in Fig. 1.23. The progress of the Call packet through the node results in the allocation of the necessary LCNs, and the mapping of these LCNs between ports via the node tables.

The first Data packet shown in Fig. 1.21 will appear as in Fig. 1.24. This will arrive at the node on port 1, and the combination of this port number and the LCN is looked up in the configuration. The mapping shows the node that the packet has to be routed out on port 3, and that the LCN field on this new link has to be 1FF. The LCN field will, in general, be modified at every node as the packet travels through the network. Packets travelling in the opposite direction are treated in exactly the same way.

During the clearing phase the configuration mapping entry is removed and the LCNs are then available for re-use by other calls.

To return to an earlier problem, if the wire breaks and layer two cannot re-establish the link, then the node can issue a Clear packet back along the network to tidy up the call. Strictly, this particular communication between layers two and three is not required by the standard although it is common. Where it is not provided then layer three is forced to use timers on its packets to detect non-delivery and take clearing action.

**Fig. 1.22** Node N LCN assignment for a new call

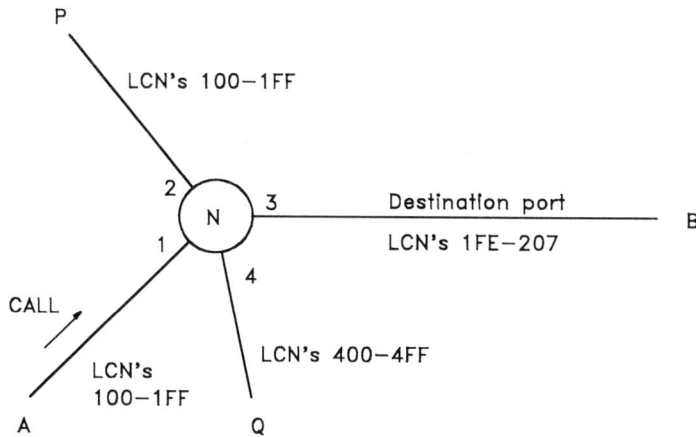

LCN's 100–1FF

P

2
N
3
1
4

Destination port
LCN's 1FE–207
B

CALL

LCN's 100–1FF

LCN's 400–4FF

A

Q

**Fig. 1.23** Node N internal tables

| Port | LCN | connected to | Port | LCN |
|------|-----|--------------|------|-----|
| 2 | 100 | | 4 | 400 |
| 2 | 101 | | 3 | 1FE |
| 1 | 100 | | 3 | 1FF |

LCN    Type    P(R)  P(S)         DATA

**Fig. 1.24** First Data packet for the call

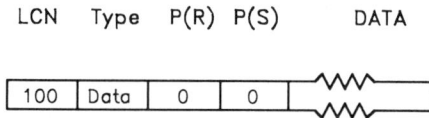

| 100 | Data | 0 | 0 | 〰〰 |

## 1.5  The network users

We have now seen how data is wrapped-up in the layer two and layer three protocols and routed safely through the network; we now have to look at what is at the end points of the network. By and large the users of the network will be people on terminals of some description, or host computers. Any combination of terminals and computers is feasible.

Clearly the users of the network – the terminals and hosts – have to interface to the packet network. This is normally shown as in Fig. 1.25. The packet network is shown as a cloud to indicate that the exact topology of the network is transparent to the two users of it.

The packet terminal can be considered to be in two parts: the interface to the user and the interface to the network. The interface to the user may be keyboard and screen, light pen, mouse or voice recognition unit.

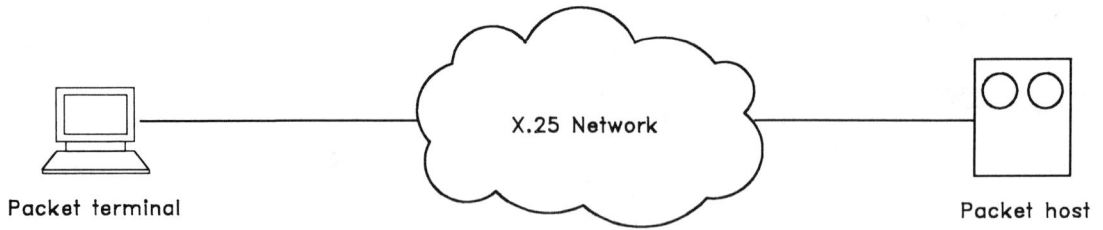

Packet terminal       X.25 Network       Packet host

**Fig. 1.25** The two ends of the call

The interface to the network takes the data from the user and wraps it into the layer two and layer three protocols. Data from the network is unwrapped from protocols and presented to the user.

This is an expensive method of connecting the user to the network:

- Network connections have to be purchased, either from a third party or by the provision of ports on the network equipment.
- The user has an individual network interface.

**Fig. 1.26** Several users sharing the same network interface

A more cost-effective way of connecting users to the network is for several of them to have the same interface and network connections, thus reducing costs. See Fig. 1.26.

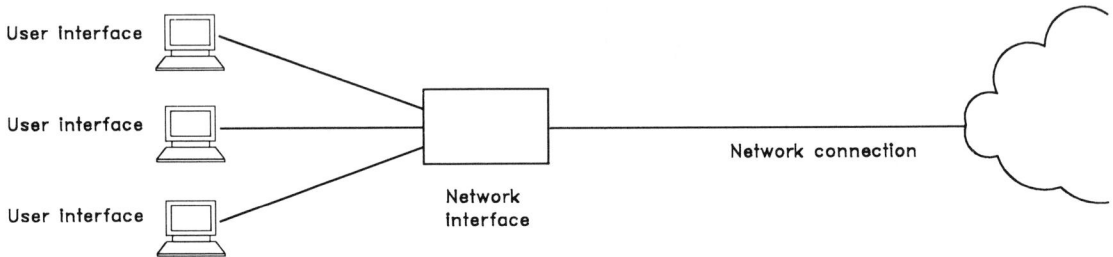

User Interface

User Interface

User Interface

Network interface

Network connection

The function of the network interface is to assemble data from the several user interfaces into packets, and to disassemble packets coming from the network into data for the user. It is commonly called a PAD (*Packet Assembler Disassembler*) and it is the Communication Processor shown on Fig. 1.3 from the early part of this chapter.

Putting together the functions outlined earlier for the CP, and what we have seen of the packet network, we can now outline much more precisely what happens when a user connects to a host service across the network:

- The user first of all makes contact with the PAD and will get a command line prompt. At this stage the conversation is purely local between the user and the PAD.
- The user will instruct the PAD to make a connection to the desired host service.
- The PAD issues a Call packet across the network and hopefully gets a Virtual Circuit established.
- The PAD would normally display a message to the user indicating successful connection.

- The host service across the network knows that a call is in progress because it has received the Call packet and issued the CAA packet. It would normally send a welcome announcement to the user automatically, in the same way that it would for a locally connected terminal.
- The PAD is now essentially transparent; data from the user will be packetized and sent into the network. Packets from the network will be disassembled and the data delivered to the user.
- At the end of the call one of three things may happen:
  1) The host service will know that the call is finished perhaps by a special transaction or the user typing "logoff". The host will then issue a Clear packet.
  2) There will be a fault in the network and one of the nodes will issue a Clear packet.
  3) The user will want to finish the call. This must be done by first of all attracting the attention of the PAD by the use of some escape sequence, and then issuing a command which causes the PAD to issue a Clear packet.
- Whichever happens, the Virtual Circuit will be cleared down by the exchange of CLR and CLC packets, and the PAD will normally give the user a message to this effect.

The PAD performs these functions for all users connected to it. This will typically be around eight but may be as many as 32. Each user is completely independent and simply shares the network line but on an individual LCN. The PAD is perhaps best viewed as several separate PADs with a concentration for the packets as shown in Fig. 1.27. For this reason the PAD is often known as a *terminal concentrator*.

**Fig. 1.27** The PAD may be considered as several individual PADs and a concentrator

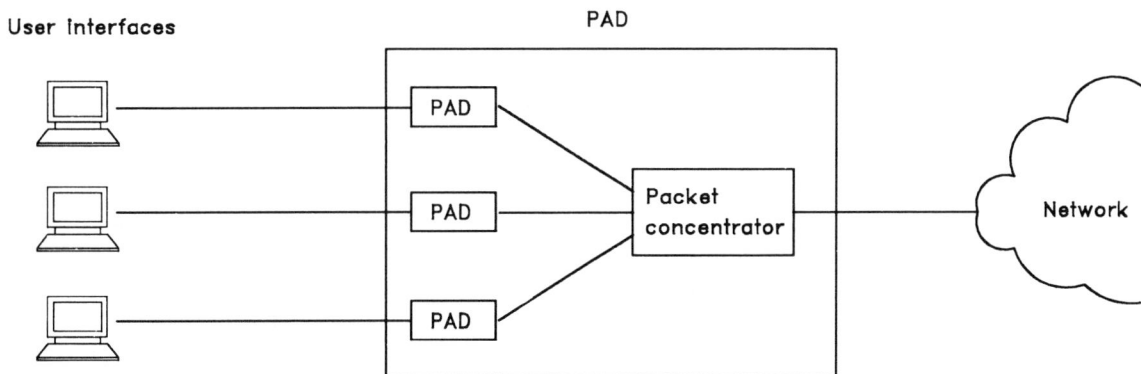

It must be remembered that the user interface may not be a straightforward terminal but may be any device capable of sending or receiving characters. If the devices are not capable of interaction with the PAD, such as light pens, then the PAD may be configured to call a preselected host automatically when the device is activated. Similarly if the user does not have any knowledge of the network and always wants to connect to the same host then a similar scheme can be adopted.

Advanced features such as these are not required by the standards but are provided by some manufacturers. Alternatively a Permanent Virtual Circuit can be allocated so that a Call Setup is not required and the network simply provides a pipe for data between the PAD and the host.

## 1.6   PAD location

The PAD may be in one of two locations depending on political and financial considerations. This is shown in Fig. 1.28.

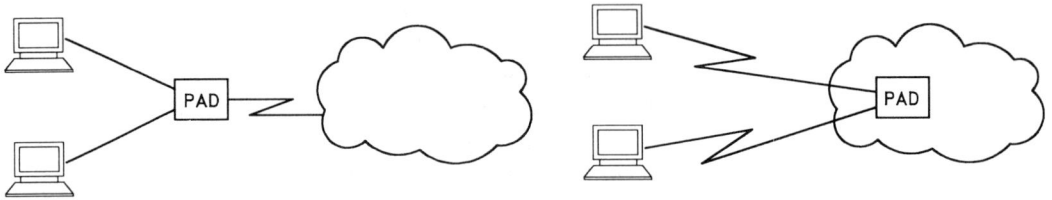

**Fig. 1.28**   Possible PAD locations

The first alternative is for a group of co-located users to have a local PAD and to share an X.25 link into the network. This link will normally be over a long distance and extra devices (modems) are required to achieve the connection. Modems are described in Chapter 3. Alternatively, the PAD may be provided by the network supplier, in which case users must each connect to the PAD via modems from the user interface.

Clearly there is no functional difference between the alternatives. The decision will be based on the charges made by the network supplier, what is provided as part of the service, and the capital cost of the rest of the equipment which has to be supplied by the network user.

## 1.7   Host interface

The host interface must perform a broadly similar function to the PAD. On one side there is the network with its requirements for packets and frames, and on the other side there are application programs that generally require straightforward characters.

Different manufacturers approach the problem in different ways. However, there is a broad standardization on the use of a microprocessor board which at least handles layer two. The interrupt rate on the main computer of handling layer two may be significant and affect the ability of the host to do its normal job. It used to be common for layer three to be dealt with by the main computer processor, but the emphasis will inevitably shift to handling this on the communications board as well. This places the whole network load on the specialized board and leaves the main processor free to do the job it was purchased for.

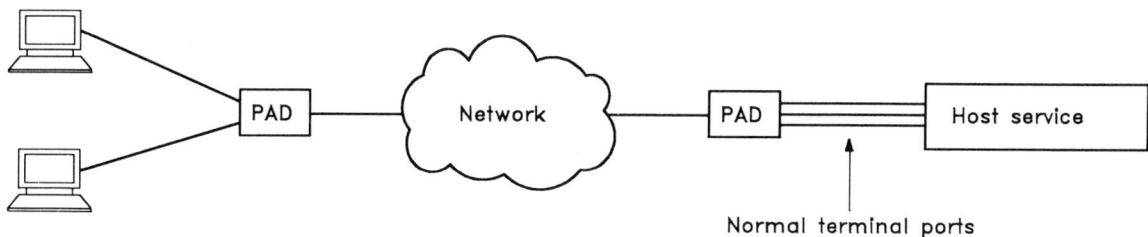

Normal terminal ports

Fig. 1.29  The Reverse
PAD

There is an alternative approach for hosts without dedicated interfaces as shown in Fig. 1.29 which utilizes the ordinary terminal ports of
the host computer, and a PAD. This system is called Reverse PAD or
Inverse PAD for obvious reasons, and the PAD is subtly different to the
terminal PAD as will be shown later. Again, a manufacturer is not
obliged to provide this feature in a PAD.

This system is not very efficient since there are two sets of terminal
ports interfacing between the host and the Reverse PAD. Reverse PAD
is generally only used as an interim solution when a network is being set
up or the full host interface is not available. Nevertheless, it does allow
all hosts to become services on a network, and provides a choice of hosts
for users to call.

# 2  The PAD and the switch

## 2.1  Introduction

Chapter 1 showed the packet switching network with particular reference to X.25, and established the need for a multi-layer approach. The implementation of the lower layers of the ISO OSI seven-layer model by X.25 was described, and some of the frames and packets were described. The need for a PAD was shown as the usual way of interfacing between the characters of a user interface and the packets of the network.

This chapter will look at the PAD in much more detail and will describe many of the features that go to make up this all-important component of the network.

## 2.2  The two ends of the call

There are three entities that the PAD has to consider at all times:

*the user      the network      the host service*

The requirements of the network are largely satisfied by the layer two and layer three protocols, and whilst some configuration is required, as will be seen later, there is nothing that needs to be changed dynamically during the call.

This leaves the two ends of the conversation with the PAD between them. The PAD must present data to each entity in a suitable form; some examples of the tasks are as follows:

- If the user screen is 80 characters wide then what should the PAD do if the host sends 81 characters without a new line?
- What should the PAD do if the user sends characters with incorrect parity?
- How does the PAD send characters from the user to the host:
    *a*) Every character is sent as an individual packet which imposes heavy loads on the host and therefore costs to the user.
    *b*) Characters are buffered and sent when the user hits carriage return.
    *c*) Characters are buffered and sent every 38th character.
    *d*) Characters are buffered and sent every 20 milliseconds.

- Should the PAD set the parity on characters going to the host?
- What if the user device is a graphics terminal and needs to swap between normal characters mode and graphics mode?

It is not possible for all of these questions to have predetermined answers given the profusion of computing devices with different characteristics. The PAD therefore needs to have a number of parameter settings which determine how it acts. These parameters are in fact dynamic and can change the way the PAD acts from moment to moment.

### 2.2.1 PAD parameters and triple-X

The parameters are standardized in the CCITT X.3 recommendation, so that all PADs are essentially the same. The following list shows some of the X.3 parameters; the full list is given at the end of this chapter. All the parameters are numeric and can take values from 0 to 255. Some of the parameters are simple two-way switches, and for these a value of zero means OFF or FALSE, and a value of a 1 means ON or TRUE.

| Parameter | Name | Meaning |
|-----------|------|---------|
| 2 | **Echo** | If 1, then the PAD echoes keyboard characters back to the screen. If 0, then characters are not echoed. This parameter would be set to 0 if the host were going to echo characters. |
| 3 | **Forward** | This defines when keyboard characters buffered by the PAD are forwarded into the network. For example, a value of 2 means forward the contents of the buffer when Carriage Return is received. |
| 9 | **Padding** | This defines a delay that the PAD should make following the sending of a Carriage Return. It is intended for use with slow devices such as printers and printing terminals. |
| 10 | **Line fold** | This informs the PAD of the width of the screen and indicates where the PAD should insert newlines into the data stream. |
| 13 | **Linefeed** | This defines whether the PAD should insert a Linefeed character after a Carriage Return either from the user or from the host, or for data echoed locally by the PAD. |

| 14 | LF Pad | This defines a delay that the PAD should make following the sending of a Linefeed. As with parameter 9, it is intended for slow printing devices. |
| 15 | Edit | Allows local editing of data in the PAD buffer by the following characters: |
| 16 | Ch del | Defines the character which deletes characters from the buffer. |
| 17 | Buf del | Defines the character which deletes everything from the buffer. |
| 18 | Buf echo | Defines the character which displays the current contents of the buffer. |

The X.3 standard defines the values allowed for the parameters and what their meaning is. These are listed in more detail at the end of this chapter.

As well as having a set of parameters, some way of setting and manipulating them is required. In fact, two ways of setting them are needed:

- The user needs to be able to set them − perhaps to inform the PAD of the width of the screen.
- The host needs to be able to set them − perhaps to set the forwarding conditions to the needs of the application program.

Each of these requirements is satisfied by separate standards:

- X.28 defines a user interface to the parameters; in fact it defines a complete set of commands for controlling the PAD including the making of calls and clearing them down.
- X.29 defines a method for the host to alter the parameters in the PAD.

The combination of the X.3 parameters, and X.28 and X.29 to control them, gives a complete method of controlling the call made by the PAD. This set of recommendations is commonly referred to as triple-X or XXX, and PADs which conform are called triple-X PADs.

Chapter 1 showed that X.28 requires some sort of escape sequence to indicate to the PAD that the next data coming from the user is a PAD command rather than data to be packetized and sent to the host. This is in fact defined in X.3 to be the *Data Link Escape* character (DLE) which is obtained on a keyboard by control-shift P.

X.29 also requires some type of escape mechanism to indicate to the PAD that the incoming data is a PAD command, not data to be sent to the user. This is accomplished by use of the Q bit in the data packet.

- If the Q bit is set to 1 then this is a qualified data packet and the data portion contains a command for the PAD. The host may request information from the PAD and this is returned by the PAD in a qualified data packet.

● If the Q bit is set to 0 then this is a normal data packet and the data is to be routed to the user or host depending on which direction it is going.

The three triple-X recommendations were made by CCITT in 1980. In 1984 they were modified slightly, along with X.25, to implement improvements and new features. A network and its attached devices will generally operate either entirely 1980 or entirely 1984 protocols. The following descriptions of the triple-X recommendations refer to the 1980 versions, and a section at the end of the chapter summarizes the main differences in the 1984 versions.

### 2.2.2   X.28 commands and service signals

The X.28 standard defines commands that the PAD user can give to control the PAD, and the service signals that the PAD displays to the user indicating what is going on. The full list of these commands and signals is shown at the end of the chapter, and the following subset illustrates the types of definitions.

*User commands to PAD*

**PAR?**    Displays the current settings of the X.3 parameters on the screen. The command can be modified to show only selected parameters – PAR?1,3,6 for example.

**SET?**    Sets the value of all of the X.3 parameters to a PAD-defined default. The command is also used to set desired values; for example to set parameter 1 to 1 and parameter 3 to 4 the command would be SET?1:1,3:4.

**CLR**    Instructs the PAD to clear the call down.

**STAT**    This causes status information to be displayed such as how long the call has been in progress and how many packets have been sent.

**xxxx**    Makes a call to the given network address. Many PADs also provide a more explicit command such as CA xxxx.

*PAD service signals to user*

**PAR2:1,3:2,64:INV**    This is the response to a PAR? or SET? command and shows the values of the relevant parameters. If one is invalid then this is shown too.

**COM**    Indicates that the call has connected.

**ERROR**    Indicates that the previous command was not valid.

A short session might therefore proceed as follows:

- User types 51243875 to cause the PAD to make a call to the given network address.
- PAD displays COM to indicate that the connection was successful. This would be followed by welcome information from the called service, and all further data transfer would be between the service and the user and the PAD would effectively be transparent.
- The user types the DLE character (control shift P) to indicate that a PAD command follows.
- The user types SET?3:2 to set the forwarding condition to be Carriage Return.
- The PAD responds with PAR3:2.
- The user types Carriage Return to resume the conversation with the service but with the new parameter value.
- The user types the DLE character to indicate that a PAD command follows.
- The user types CLR to clear the call down.

If the user wanted to send the DLE character to the host computer then X.3 parameter 1 must be set to 0 to indicate to the PAD that no interception is required. A rather more concise way is for the user to enter two DLE characters – the second one is then effectively a command to the PAD to insert a DLE character into the input buffer and to return to normal character input.

### 2.2.3  X.29 commands and responses

We saw above that X.29 commands and responses are carried in layer three packets with the Q bit on. A typical packet is shown in Fig. 2.1.

**Fig. 2.1** Conceptual layout of the packet showing the Q bit

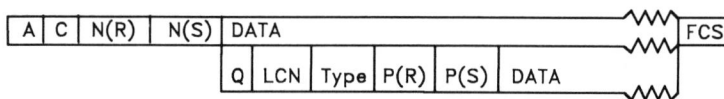

The following list shows the X.29 command types. The values are normally shown as two hexadecimal digits to avoid possible confusion over how the value is encoded in the byte.

| First byte | Name | Explanation |
|---|---|---|
| 02 | **Set param** | This is a message from the host to the PAD to set a list of X.3 parameters. |

The second and third bytes will normally be a parameter number and the desired value respectively. Subsequent pairs of bytes may also be provided if several parameters are to be set. If no parameter pairs are provided then the parameters are all set to default values.

**04**          **Read param**    This is a message from the host requesting a listing of the values of a list of parameters.

Subsequent bytes contain the parameter numbers for which values are required. The PAD will respond with a parameter indication message.

**00**          **Param ind**    This is the parameter indication message from the PAD to the host showing how X.3 parameters are set.

Pairs of subsequent bytes contain parameter numbers and current values. Only the parameters requested in a read parameters message are normally returned.

**06**          **Set & read**    Combined set and read parameters.

This message from the host has the same format as the set parameters message and does the same job. However, it also requests confirmation from the PAD that the settings have been made. The PAD will respond with a parameter indication message.

**01**          **Inv to clr**    This is a message that indicates the end of a call. It is not as destructive as a Clr packet.

The Invitation to Clear message is most often sent by the host computer at the end of a session. The layer three Clear packet is not part of the actual data stream and therefore does not contain N(R) and N(S) numbers, and often overtakes data in the network that is held up waiting for a window to open. This means that, if the host sends a Clear immediately after any data, then the Virtual Circuit may be cleared down before the data is delivered. The data is then flushed and the user will not receive it. By sending this X.29 control message – which must necessarily be in a data packet following the normal flow control mechanism – the host can be sure that the data is all delivered before the PAD receives the Invitation to Clear. The PAD responds to the Invitation to Clear by issuing a Clear packet to clear the call.

**03**          **Ind of brk**    Indication of break.

This is a message from the PAD to the host indicating that the user has requested a break in output, perhaps in the middle of a file listing. The host will normally stop sending data on receipt of this. The second and third bytes may optionally contain values of 8 and 1 respectively, indicating the current setting of parameter 8. This is used with parameter 7 as explained in the full X.3 list of parameters.

**05**          **Error**    Indicates that the last message received was invalid in some way.

### 2.2.4 Problems with triple-X

There are a number of problems with the triple-X mode of working:

- There is no responsibility assigned for which end will set the parameters, only the ability to set parameters as required. This means that a particular user who accesses several hosts may have to perform a different level of parameter setting for each if the hosts are different.
- The opposite problem is that both the user and the host may assume responsibility for setting parameters, and thus cause a great deal of confusion.
- There is no standardization in methods of using triple-X. Thus different host manufacturers, different application programmers and different users may all do different things to bring about the same effect. For example, to achieve line-at-a-time operation:
  1) The default parameters may be correct so neither end needs to do anything.
  2) Host A may set forwarding on alphanumeric characters (P3 = 1) as the first and only parameter setting it does, so the user needs subsequently to change P3 to 2.
  3) On host B every application program may set appropriate parameters, so the user does not need to do anything.
- X.29 can only be used on X.25 networks and never on other types of packet switching network because of the use of the Q bit.
- X.28 is not very friendly to novice users.

Despite these problems triple-X is still the international standard for X.25 PADs and is adopted on the majority of products.

One alternative was proposed by the PSS User Forum, a body of users of the British PTT. This proposal on the Use of Character Protocols on PSS was made in a document known as the "Green Book" (named from the colour of the cover). Amongst the points made in this document are:

- If a set of users have their own PAD then there is no reason why they should use X.28 to give commands and receive service messages. Any system may be chosen as long as it does not affect the "network side" of the PAD.
- It assigns responsibility for PAD parameters to one end or the other. Thus, for example, the width of the terminal (parameter 10) is set by the terminal end via X.28 – or whatever system is chosen – and should not be set by the host end.
- It recognizes that there are a limited number of settings of the parameters that any application might need, and suggests three standard operation modes.

The three Green Book operating modes are shown below.

### Message mode

> Forward on CR ETX EOT
> Local editing allowed
> PAD has control of output

This implements line-at-a-time operation. The user assembles a line of data locally and sends it in a single message to the host. Data from the host is processed by the PAD to ensure that it is suitable for the terminal. For instance, linefeed may be added (parameter 13), the line may be folded to fit the screen (parameter 10), and padding may be inserted (parameters 9 and 14).

Message mode is suitable for applications that do not echo data and process the data only on receipt of a Carriage Return. The PAD can therefore perform local editing of data before it is sent to the host. It is generally efficient in that it minimizes the number of packets sent across the network.

### Native mode

> Forward on timeout of 1/20 sec.
> PAD does not process output

For most people this implements character-at-a-time operation where, basically, every character from the user is sent in its own packet to the host. It is therefore expensive as it imposes a considerable network load when the amount of data rises. Since the forward condition is based on a timer then very fast typists can get characters aggregated into packets. Generally this advantage is only experienced by automatic devices such as pre-programmed function keys or auto-repeat features of terminals.

In this mode the PAD performs no processing of data from the host, so the host has total responsibility for formatting the user's screen. Note that this is a departure from triple-X since the PAD does not act according to the X.3 parameters.

Native mode is generally used for applications that echo data immediately it is received, or in graphics applications where the data is not character-oriented, or in applications such as screen editors where single characters are commands that cause radical changes to what is on the screen.

### Transparent mode

> Forward on CR ETX EOT
> Local editing allowed
> PAD does not process output

This allows Message mode in the forward direction from the user, but leaves the host to control the screen. It is therefore useful in some graphics applications where commands can be assembled locally but the response is not character-oriented.

Despite early popularity Green Book operation has largely died out, and nowadays it is only the user-friendly alternative to X.28 that is useful. Even that is not too attractive in the climate of international standardization. A similar type of work is being carried out by CEN/CENELEC – the Joint European Standards Institution – as ENV 41901, or, as it is usually called, Y11–Y12. This recommends five standard sets of parameters with assigned responsibilities. However, it works within X.28 and X.29.

## 2.3  Reverse PAD

The use of Reverse PAD to provide the host-end connection to the network was introduced in Chapter 1. We can now see a little more of the detail of what is required for the implementation. Fig. 2.2 shows the network arrangement.

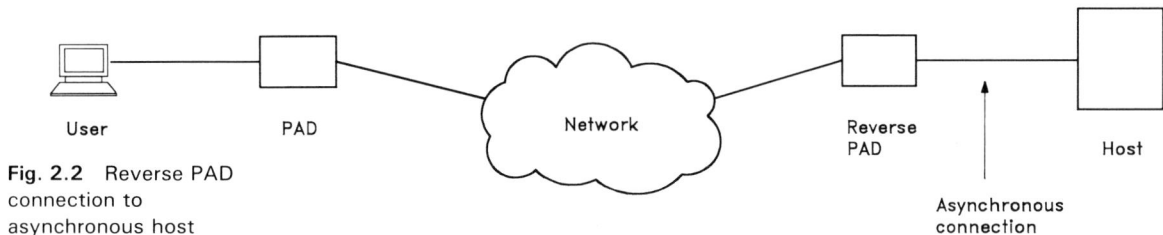

**Fig. 2.2** Reverse PAD connection to asynchronous host

The first point to make is that the triple-X standard is still adhered to. The X.3 parameters are located in the user PAD, and the Reverse PAD has a completely different set of parameters to describe the connection with the host. The Reverse PAD will also need to provide an X.29 host-end implementation to give commands to the user PAD. This is a necessity because the PAD may send an X.29 indication of break message that the Reverse PAD must deal with.

The most significant question in configuring the Reverse PAD is whether the host is going to echo every character.

- If it is, then the system has to forward characters from the keyboard, through the user PAD, across the network, through the Reverse PAD and into the host, and then get the echo back and on the screen as quickly as possible. This means that the Reverse PAD has to send an X.29 Set Parameters message at the start of the session – after it has sent the Call Accept – to set the user PAD into forwarding each character immediately.
- If the host does not echo characters then no particular action is required by the Reverse PAD since anything that the user PAD is configured with as a default will be acceptable. It is probably tidier though for the Reverse PAD to set the user PAD to forward on Carriage Return, and perhaps on a timer too, to ensure that data does actually arrive.

In either case, data from the host must all be forwarded immediately across the network to the user, and the Reverse PAD will generally operate a system of timeouts so that data is not delayed, but when there is a lot of data it is aggregated to use the network more efficiently.

It is theoretically possible to allow the host to communicate with the Reverse PAD, along the lines of X.28, to alter the way in which data is forwarded by the Reverse PAD, or to set the user PAD for the application. This would then be similar to the way in which a human user interacts with a PAD to set parameters for particular circumstances. Such a system is difficult because of the lack of international standards, and in any event it is probable that if that amount of complexity were possible in the host then it would be possible to implement X.25 properly.

Overall, it is probably better to keep the host interface as simple as possible. The host then does not need to know it is connected to anything other than a dumb terminal, and there is no need for another protocol.

## 2.4   PAD commands

We have already seen the X.28 commands that allow a user to establish and control a call. The PAD must also have a set of configuration commands that allow the site manager to put the PAD into service.

### 2.4.1   Port configuration

A PAD may have three or more types of port on it, including X.25 network port, user interface port, and host port for Reverse PAD. The number and mix of these types is dependent on which manufacturer and which product is chosen. Configuration of the ports is carried out in two phases: physical setting of jumpers and switches (if any) to configure the hardware interface; and logical setting of software parameters

**Hardware**   The amount of electronics that needs to be configured varies between none at all if the ports each have only a single function, and a considerable amount if the ports may be configured for a number of functions. The electronic aspects are considered in Chapter 7, but clearly every PAD has its own set of switches and jumpers and it will be necessary to refer to the manufacturers' literature.

There is a move away from setting up the hardware using switches and jumpers, to a much more convenient setting by software commands. The commands cause values to be set in configuration registers, and the registers then cause the hardware to be set in the same way as if a physical switch had been used.

Some of the larger integrated circuits that make up the hardware have this type of feature already built-in. For example, Baud Rate Generators have their speed programmed from the data bus so in most PADs the speed of the lines is set from a software command. There are some PADs which require no physical hardware setting at all due to this technique, but most PADs will use it to a certain extent. The site manager can expect to set the speed of the line, the parity, the number of stop bits, and the flow control characteristics by system commands.

**Software**   Software configurations of a port will probably proceed in two phases: setting of the hardware as explained above, and logical configuration. The logical configuration will depend on the type of the port:

- *Asynchronous user ports*: the connection is expected to be to a terminal, so the configuration will describe the terminal to the PAD. This is essentially the user's half of the X.3 parameters, and the configuration will become the parameters whenever the user makes a call. This then relieves the user of having to explicitly describe a terminal by X.28 commands.
- *Asynchronous Reverse PAD ports*: the connection is to a terminal port of a host computer so the PAD has to act as a terminal. This requires no configuration other than for the hardware. What must be done though is to describe the way that the connection is to operate as discussed earlier in this chapter.
- *Synchronous X.25 ports*: the connection is to a network and the main aspect of configuration is which logical channels are to be used.

Typical conversations between the site manager and the PAD are shown in Fig. 2.3 and Figs. 2.4 and 2.5.

```
SET LINE

which line?    7
what type?     USER
speed?         9600
parity?        EVEN
stop bits?     1
flow?          XON/XOFF
width?         80
...
```

**Fig. 2.3**  Port configuration for terminal user lines (PAD configuration questions are shown in lower case and the responses of the site manager in upper case)

```
SET LINE

which line?    3
what type?     REVERSEPAD
speed?         9600
parity?        ODD
stop bits?     1
flow?          XON/XOFF
echoing?       YES
```

**Fig. 2.4**  Port configurations for Reverse PAD lines (PAD configuration questions are shown in lower case and the responses of the site manager in upper case)

**Fig. 2.5** Port
configurations for X.25
network lines (PAD
configuration questions are
shown in lower case and
the responses of the site
manager in upper case)

```
SET LINE

which line?      2
what type?       X.25
speed?           2400
LCNs?            400—4FF
```

A popular option is to configure terminal characteristics in a *Profile*, so that the PAD has knowledge of all possible hardware that may connect to a user port. The Profile can then be named when configuring ports, thus easing the task of repeating the same description for many ports. Also, the user can adopt a new Profile very easily if a different terminal is to be connected to a port. By giving a command such as Profile ABX3, all the characteristics of the ABX3 terminal are loaded into the X.3 parameters without the need to give individual X.28 commands for every parameter.

The X.28 Profile command performs this function but also sets the X.3 parameters relevant to the host as well.

### 2.4.2 Address configuration

The X.28 commands provide the ability for a user to make a call out onto the network by giving the address to which a connection is required. We saw in Chapter 1 that the address is conceptually like a telephone number, and that it may be 15 digits long.

One of the useful features that a PAD may have is a *mnemonic address* mechanism that allows keywords to be used instead of the actual address. Thus if the PAD has been configured with the knowledge that WORDPROC is at address 5813421764791 then the users of the PAD can use the command:

CA WORDPROC
instead of CA 5813421764791

Such a feature makes the PAD non-compliant with X.28.

It is necessary for a definite command to be used to make the call when using mnemonic addresses, otherwise there could be confusion over whether a mnemonic address or a command was being given. This clearly makes life easier for the users. There are a number of mechanisms to implement this, but essentially they all rely on a table, and may be configured as shown in Fig. 2.6.

There are some problem areas with this type of scheme. Firstly, there needs to be a known search order of mnemonic addresses in the table: is it alphabetic, is it by length, or is the search order configured by the site manager when the table is created? Secondly, there has to be a known

**Fig. 2.6** Configuration of mnemonic address translation (PAD configuration questions are shown in lower case and the responses of the site manager in upper case)

```
SET TRANSLATE

keyword?        WORDPROC
address?        5813421764791
```

**Fig. 2.7** Address translation with host X.3 parameter defaults (PAD configuration questions are shown in lower case and the responses of the site manager in upper case)

```
SET TRANSLATE

keyword?        WORDPROC
address?        5813421764791
forward?        2              (parameter 3)
timeout?        0              (parameter 4)
editing?        YES            (parameter 15)
...
```

method of dealing with keywords that do not match exactly: for instance if WP is the table, what happens if the user gives a command CA WP7?

An advantage of having a table is that, because the site manager knows in advance which hosts are going to be called, the host's half of the X.3 parameters can be configured as was done in the line configuration for the user. This means that as soon as the call is connected there is a complete and consistent set of X.3 parameter settings. The configuration may be something like that in Fig. 2.7.

### 2.4.3  Text messages

Text messages from the PAD to the user are defined in X.28. However, as with address configuration there are some useful extra configurations that help the user, but which again make the PAD noncompliant with X.28. These are all dependent on the specific PAD in use.

*Announcement text*  This is a block of text displayed when the user first activates the terminal line. It will usually give items of news, and may also list the mnemonic addresses held in the address configuration table and known Profiles. It could also list some or all of the available commands.

*Information text*  This is text that gives the user information when in difficulty. It is usually hierarchical so that at the first level all the available commands are listed. At the second level the user asks for information on a specific command and is given all the options that are possible. At the third level full details are given.

*Clearing reasons*   When a call is cleared from the network side — that is, not cleared by the user — then a clearing cause code is contained in the Clr packet sent to the PAD. The PAD will normally just display this code on the screen; however it is possible to give a more friendly message as long as it is known in advance what the codes mean. Some of the codes are specific to the network and are allocated by the network administrator, but generally the same codes are used for all networks as shown in Appendix B.

Configuration of the messages is again normally done via a table mechanism so that the codes are looked-up when the Clr arrives, and the appropriate text is then displayed to the user.

### 2.4.4   Internal statistics generation

The PAD may maintain counters and registers of the various activities that it performs, and these can be displayed and manipulated by the site manager. The type of statistics and their interpretation is discussed in Chapter 5.

### 2.4.5   General control

There may be a number of features controlling the overall PAD activities and the activities of individual lines. These will vary a lot between different PADs but the following features are common:

- *Configuration control*: the ability to remove configuration data and return to a manufacturers' default, or to operate several independent configurations and switch between them. This latter feature allows test setups to be tried-out without losing the known good configuration, and allows one PAD to act as a backup to a more important one.
- *Software control*: the ability to start and stop overall PAD operation. This is of most use in re-booting the PAD after a problem, or perhaps after reconfiguration.
- *Stop an individual line*: this causes a line to cease operation either because it has a fault and is affecting PAD operation, or it needs to be taken out of service for reconfiguration, or there is some faulty equipment on the line affecting PAD operation. It may also be used on terminal lines to prevent access to the network.
- *Start an individual line*: this may be necessary after the situation that brought about the stopping of a line has cleared. Perhaps the faulty terminal has been replaced or the configuration has been performed. It may also be used to clear a "deadly embrace" where both ends of a link have got into a state where they are waiting for the other to do something.

● *Controlled stop of an individual line*: this is another form of stopping a line, but rather than take the line out of service immediately the PAD waits until all current activity has stopped. If this were applied to an X.25 port, for example, the PAD would wait until all calls had cleared before stopping the line. It is clearly the most friendly way to users of taking a port out of service. The PAD may or may not allow new activity on the port during the period before the stop actually happens. This will depend on the implementation or may be an option that the site manager can exercise.

## 2.5 Call redirection and call reestablishment

Consider the situation shown in Fig. 2.8 where an X.25 call is active between two nodes A and B. If the link between A and B now goes down then what can be done to reroute the call via node C? The unfortunate answer is that using X.25 there is practically nothing. One of the nodes, say A, would have to make a call via C to B and receive the accept back,

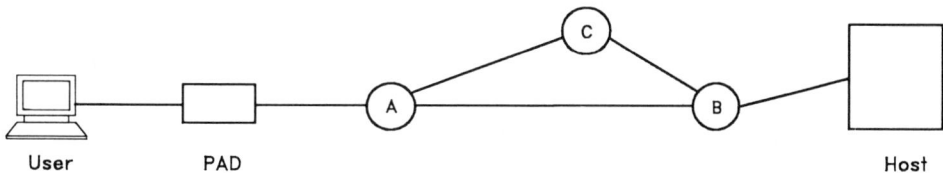

**Fig. 2.8** Example network with two possible routes between user and host

then both A and B would have to rearrange their LCN tables to recreate the circuit before resetting the call and retransmitting the lost packets. This process is simply not defined in X.25 and can only be accomplished by a higher-level protocol. Such a protocol could be used transparently within the network, but in general a break will cause the complete virtual circuit to clear down.

Once the call has cleared, then if it is reestablished in some way, then it is quite in order for it to be routed via node C if the A−B link is still down. This redirection of the call can be performed automatically by the network node. The node needs to try the first route and fail, then examine the cause of the failure to determine that it was due to a network fault, rather than, say, a normal Clear issued by a host to end a user session. The node can then try a different route. The use of alternative routes needs to be configured in the addressing tables of the nodes in the network; and is illustrated later in this chapter

Alternatives may also be configured at the PAD initiating the call, and in this case there is a choice of whether to try a different route to the same service, or to use a different service altogether. The use of the alternatives is activated by the receipt of a Clear rather than an Accept to the initial call.

## 2.6 Switch configuration

An X.25 switch, or node, is simply a unit with several X.25 ports whose function is to route a call packet in a network and establish the virtual circuit. This was discussed in Chapter 1.

Since the switch only has X.25 ports then there is only one type of port to configure, and the options will be as illustrated for the X.25 ports of a PAD. Since lines on a switch are more likely to be used on the trunk routes of a network than lines on a PAD, then it is likely that higher speeds will be offered to cope with the greater amount of traffic.

**Fig. 2.9** Switch address configuration (PAD configuration questions are shown in lower case and the responses of the site manager in upper case)

Address configuration of a switch is somewhat different to that of a PAD, since the basic decision is simply one of deciding which port to route the call out of. Configuration may proceed along the lines shown in Fig. 2.9 where the leading digits of the address are recognized.

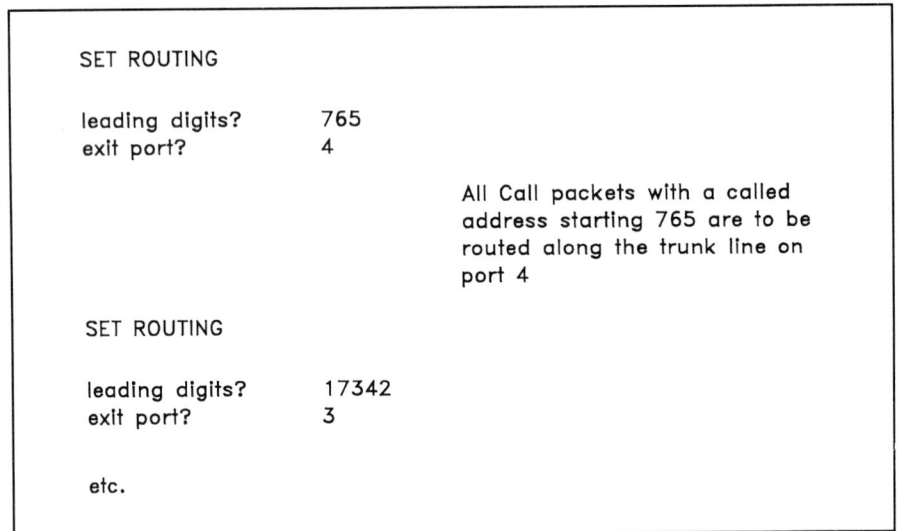

---

SET ROUTING

| leading digits? | 765 |
| exit port? | 4 |

               All Call packets with a called address starting 765 are to be routed along the trunk line on port 4

SET ROUTING

| leading digits? | 17342 |
| exit port? | 3 |

etc.

---

It is possible to get much more sophisticated than this. One enhancement is to have a different set of routings for each port. Thus calls arriving from one part of the network may be routed differently to others, even though they are destined for the same service. A further enhancement is to allow translation of the addresses much as was illustrated for the PAD. This is likely to be of more use in private networks where there is much greater freedom of address usage than in public ones where the addresses can never be changed.

**Fig. 2.10** Example network with two routes to host

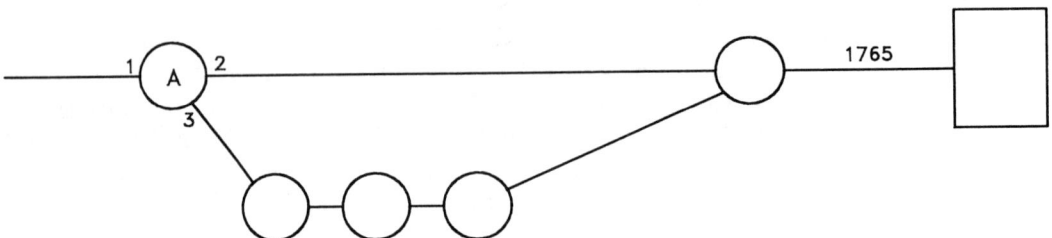

Fig. 2.11 Configuration
of alternative routing
selection in the Switch
(PAD configuration
questions are shown in
lower case and the
responses of the site
manager in upper case)

```
SET ROUTING

leading digits?        1765
exit ports?            2
alternatives?          YES
2nd exit port?         3
alternatives?          NO
```

Alternatives to routes may also be configured. Consider the network shown in Fig. 2.10 where a call is destined for a host service. The two routes may be configured as shown in Fig. 2.11.

## 2.7 Parallel routes

The situation shown in Fig. 2.10 and Fig. 2.11 allows a call, or a number of calls, to be sent on different routes if a particular route fails to get a connection achieved. Assuming both of the routes are working, then it is probably desirable for the two routes to share the calls. If a failure occurs then obviously all the new calls would use the functioning route.

Fig. 2.12 Configuration of load balancing in the Switch (PAD configuration questions are shown in lower case and the responses of the site manager in upper case)

This technique of sharing the load may be spread over any number of ports and may be configured using *balanced groups* as shown in Fig. 2.12. It is necessary to specify the loading since the ports may not be the same speed, the lines may be error prone, or may be charged on different tariffs. The switch will probably ensure that the loadings add up to 100%!

```
SET BALANCE

balance group no.?     2
exit port?             6
loading?               .3
exit port?             8
loading?               .3
exit port?             9
loading?               .4




SET ROUTING

leading digits?        31476
exit port?             #2        (identified as balance group
alternatives?          ... etc       rather than individual port)
```

In order to simplify the use of this mechanism the switch will probably treat the balanced group as a self-contained individual port, and will not allow the ports making up the group to be directly assigned in routing tables. Nor will it allow ports to belong to a number of balanced groups.

The concept of balanced groups, given these restrictions, is very straightforward. Given information on the usage of the ports within the group it is easy to see how the switch will allocate a new call. The difficulty is in how the switch measures the usage of the ports. The following factors may be used in deciding:

- Number of calls on each port – but this gives no information on the type of calls and the traffic load they generate.
- Number of packets per second – but this gives no information on the length of the packets or what type they are.
- Number of data packets – but measured over what period and how good a prediction do you get of what is likely to happen in the future?

Whatever algorithm is used it is only a guess of future loading. It is because of this uncertainty that restrictions of the type outlined above are likely to be imposed.

## 2.8 The PAD switch

There is a tendency for manufacturers to make PADs with single X.25 ports, as a completely separate product from the switch that acts as a node in a network. Even if the PAD has two X.25 ports to allow for alternative routing strategies, they are likely to be for outgoing calls only. The reasons for this separation are based on the traditional view of an X.25 network supplied by an administration, and only a small number of users at any location requiring to connect. This is illustrated in Fig. 2.13.

Whilst this model may be applicable in some cases, it is not generally representative and fails to serve some important applications:

- Where a particular site has a large number of users to connect then even if the PAD can cope with the number it is unlikely that they will all be conveniently co-located with the PAD. The solutions to this are to connect using modems or line drivers, or to use several PADs and a local switch, or to have several links in the network. These are shown in Fig. 2.14. To use modems and separate wires to each user will create a great deal of wiring and means that a significant cable run is not protected against error by protocols. Having several links to the network is likely to be very expensive in terms of annual rental from the administration. The use of a local switch provides the most attractive topology but is expensive in terms of capital.

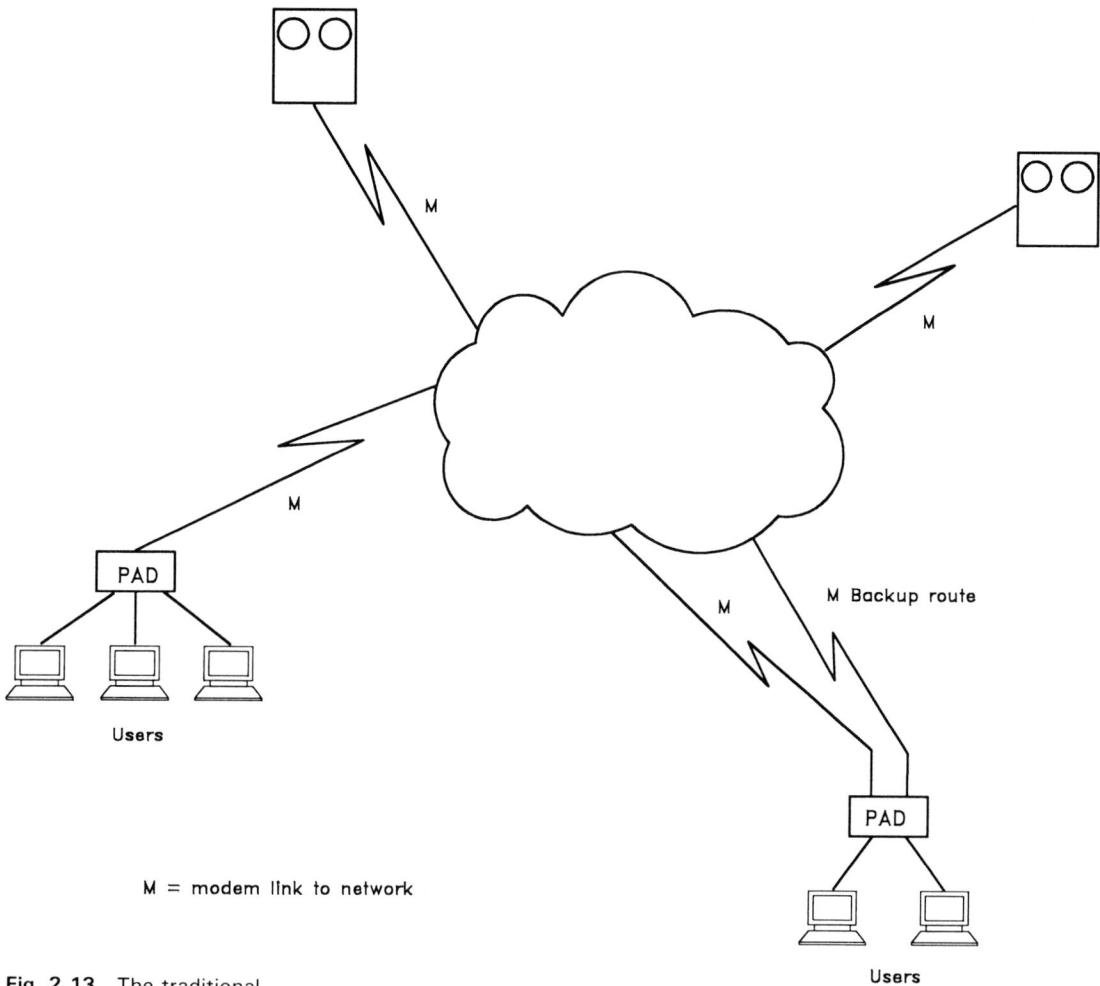

**Fig. 2.13** The traditional view of a network

M = modem link to network

● Not all networks are provided by an administration. Users may lease lines directly from the PTT and implement their own network of anything up to international scale. Here again there are likely to be small pockets of users that will require a great capital outlay to get connections. See Fig. 2.15.

One answer to these problems is a combined PAD and switch. The mix of X.25 and user ports varies with manufacturers, and ideally is configurable by the user to meet precise requirements. Fig. 2.16 shows how such a product meets the problems outlined above.

Use of a PAD switch also gives the site manager the ability to implement some alternative routing strategies as shown in the diagram and gives much more flexibility in altering the network topology at a later date.

**Fig. 2.14** Possible solutions to the problem of many users

CONNECTING MANY USERS BY USING SEVERAL LINKS TO THE NETWORK

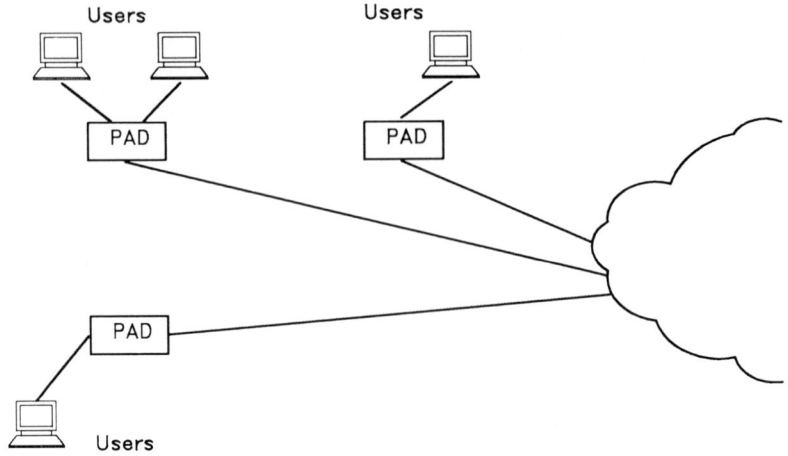

CONNECTING THE USERS BY USING A SINGLE PAD AND LINE DRIVERS

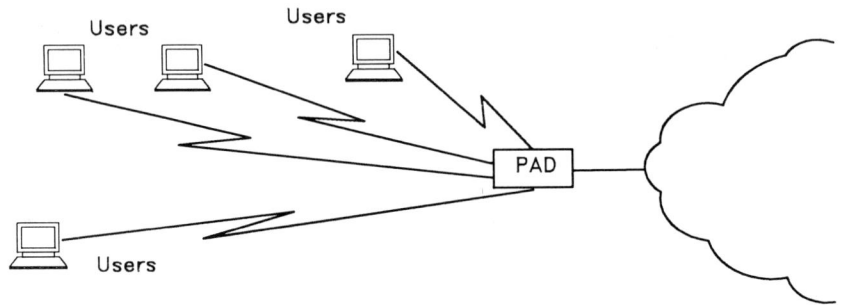

CONNECTING THE USERS BY USING SEVERAL PADS AND A LOCAL SWITCH

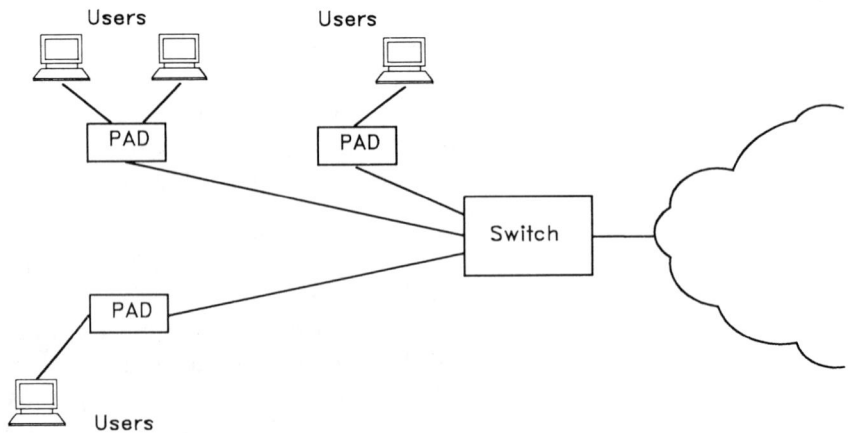

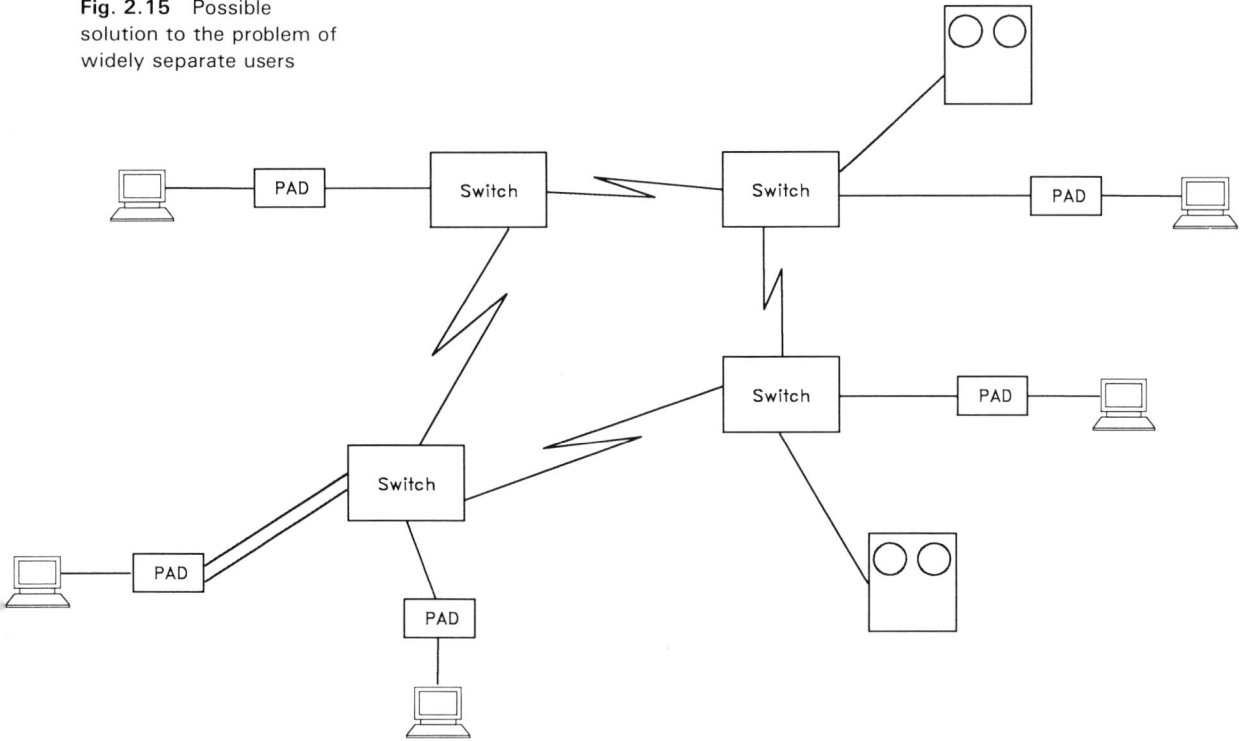

Fig. 2.15 Possible solution to the problem of widely separate users

## 2.9  X.3 (1980) parameters

The following shows the full list of X.3 parameters and their meanings. These are from the 1980 version of X.3 which is normally used with 1980 X.25, X.28 and X.29. The 1984 differences are shown at the end of the chapter.

The names shown for the parameters are not standard and are simply the usual shorthand references. The text in brackets is the standard descriptive way of referring to the parameters.

| Parameter | Name | Meaning |
|---|---|---|
| **1** | **Escape** | (PAD recall using a character.) Determines whether or not the user is allowed to escape from data transfer state to PAD command state to give commands to control the call. |

If 1, then the user is allowed to escape from data transfer state to PAD command state using the DLE character. The user would then be able to give commands to the PAD, for instance to clear the call. If 0, then no escape is allowed. A value other than 1 or 0 defines a character other than DLE for escaping, though this option is not always implemented.

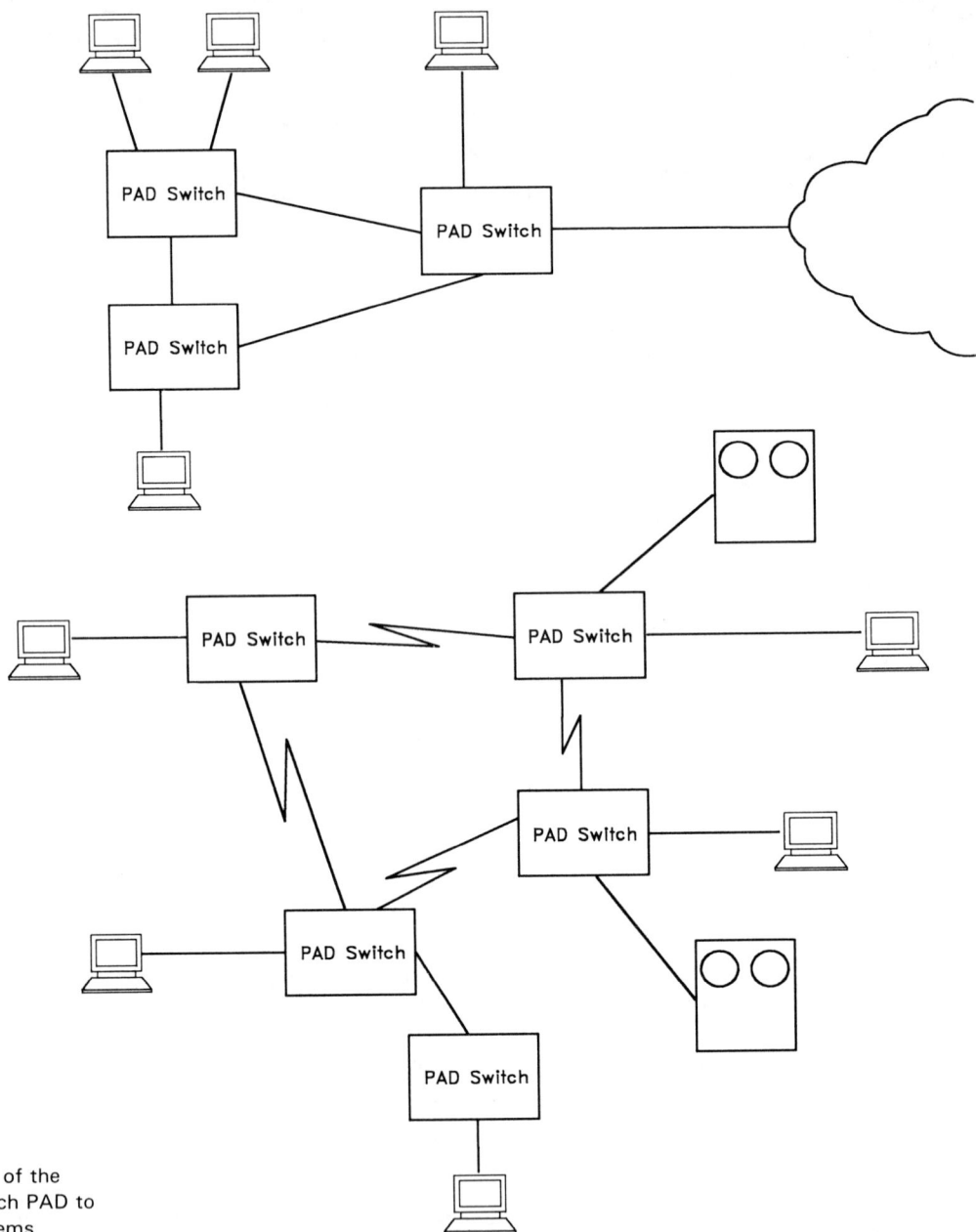

**Fig. 2.16** Use of the combined Switch PAD to solve the problems

| 2 | Echo | (Echo.) |
|---|------|---------|
|   |      | Determines whether the PAD echoes characters received from the keyboard back to the screen. |

If 1, then the PAD echoes keyboard characters back to the screen. If 0, then characters are not echoed.

| 3 | Forward | (Selection of data forwarding signals.) |
|---|---------|------------------------------------------|
|   |         | This defines when keyboard characters buffered by the PAD are forwarded into the network. |

The mechanism for collecting data from the keyboard and sending it to the host service is to accumulate characters and send them when enough data has been collected to fill a packet. This mechanism is not at all useful for most situations. Normally the application will need the PAD to send data when the user hits Carriage Return, or perhaps it will need each character to be sent individually. Parameters 3 and 4 define additional forwarding conditions to the full packet, to meet these requirements.

Parameter 3 sets a forwarding condition on characters entered by the user. For example, a value of 2 means forward the contents of the buffer when Carriage Return is received. The value is encoded in binary such that each bit represents a particular forwarding condition. The overall forwarding is then the accumulation of the bits that are set. The bits have the following meanings:

| | |
|---|---|
| bit 0 (decimal 1) | Alphanumeric characters A–Z, a–z, 0–9 |
| bit 1 (decimal 2) | Carriage Return character (CR) |
| bit 2 (decimal 4) | Control characters ESC, BEL, ENQ, ACK |
| bit 3 (decimal 8) | Control characters DEL, CAN, DC2 |
| bit 4 (decimal 16) | Control characters ETX, EOT |
| bit 5 (decimal 32) | Control characters HT, LF, VT, FF |
| bit 6 (decimal 64) | Other control characters in columns 0 and 1 of IA5 |

The characters referred to are the international standard characters used between components of a network, and which are defined in International Alphabet 5 (IA5). This is listed in Appendix C. If the application required forwarding when the user typed CR or ESC into the buffer, then parameter 3 would be set to the value 6. This is found by adding 2 (for CR) and 4 (for ESC) together. Note that forwarding would also occur on BEL, ENQ and ACK in this case, and that it is not possible in general to set the precise requirements. If the parameter is set with all bits on, then forwarding occurs on every character that enters the buffer and there is therefore no delay in getting characters from the user to the host. This may be useful if the host is echoing characters, but is expensive because data is being transferred in single-character packets. If all bits are off, that is the parameter is set to zero, then there is no forwarding condition except for full packets and what is set by parameter 4.

| 4 | Timeout | (Selection of idle timer delay.) |

This is another forwarding option that sends the contents of the buffer into the network if a period elapses during which no characters are received.

This parameter allows the buffered characters to be sent if the user stops typing characters. The value of the parameter is simply the duration of the delay in twentieths of a second, so a value of 20 means: forward the buffer if no further input is typed for one second.

This parameter might be used in preference to parameter 3 when the host is echoing characters, since it allows the user's characters to be packetized together to a greater extent. If the parameter is set to zero then the timeout does not operate at all. Parameters 3 and 4 may be used together, in which case forwarding occurs either when the character condition is reached or when the timeout is reached, whichever is first. A sensible combination of settings for the two parameters will ensure the best compromise of fitness for the application and economy in the number of packets exchanged. These parameters are additional to the packet-full forwarding which always operates.

| 5 | Control | (Ancillary device control.) |

Normally allows the PAD to control the flow of data input from the user.

The parameter may take the following values:

decimal 0   No use of flow control.
decimal 1   Flow control used in data transfer state.
decimal 2   Flow control used in both data transfer and command state.

If flow control is set, then the PAD can prevent the user sending characters by transmitting flow control characters XON and XOFF to the terminal. This only affects characters typed at the keyboard. This would usually be done where a very fast terminal or user device was attached to the PAD, and the PAD software was unable to deal with the input character rate. It may also be needed if the X.25 window is full and no more packets could be sent. In this case the PAD can choose to buffer data, but will eventually run out of space and be unable to store any more untransmitted data.

| 6 | Suppress | (Control of PAD service signals.) |

Stops the PAD sending service signals to the user.

When the parameter is set to 0 then the PAD will not send its own messages to inform the user about what is happening in the PAD and network. Other values are bit-oriented as for parameter 3. The bits have the following meanings:

bit 0 (decimal 1)      Service signals other than the prompt may be sent by the PAD.

bit 1 (decimal 2)      The prompt may be sent by the PAD.

Further bits have values dependent on the network and are used to set the language (English, French, etc.) in which the signals are given. Not all PADs implement this!

| 7 | **Break** | (Selection of operation of PAD on receipt of break signal from the start-stop DTE.) This determines the action of the PAD when the user indicates that a break is required. |

Break is a key on a terminal that causes a special condition to be raised on the cable which the PAD will always detect. It is therefore like a command given by the user to the PAD. Being a special condition, not a character, it can always be given whatever state the PAD is in. Chapter 7 explains the condition in more detail. Parameter 7 defines the actions of the PAD on receipt of this condition, and again is bit-oriented as for parameter 3. The bits have the following meanings:

bit 0 (decimal 1)      Send an Interrupt packet to the host.

bit 1 (decimal 2)      Send a Reset packet to the host.

bit 2 (decimal 4)      Send an X.29 Indication of Break to the host.

bit 3 (decimal 8)      Escape from data transfer to PAD command state.

bit 4 (decimal 16)      Discard all output buffered in the PAD awaiting transmission to the screen.

A value of zero means: do nothing. The various packet types mentioned above are explained in more detail in later chapters. Their effect is dependent on the particular host in use and, in general, will cause the host to stop what it is doing.

Bit 4 (discard output) is never used on its own and is only actioned by a value of 21 for parameter 7. If this is set and the user hits the Break key, then:

- The PAD sends an Interrupt to the host.
- The PAD sets parameter 8 and discards all data that has already been received from the host, or that is received in the future.

● The PAD sends an Indication of Break message, that also carries a parameter indication message informing the host that parameter 8 is now set to 1.

This combination of actions ensures that output to the screen stops as soon as the user hits the key, and the PAD and the host between them "tidy-up" data queued in the network.

| 8 | **Suppress** | (Discard output.) This prevents the PAD delivering data to the screen. |

This would be set to 1 by the PAD when the user had requested—via the Break key and parameter 7 set to 21—that transmission from the host should stop. It causes the PAD to discard all output buffered in the PAD awaiting transmission to the screen, and also applies to any screen data subsequently received by the PAD. Output on the screen therefore stops immediately. On receipt of the interrupt and X.29 message, the host would stop sending data to the PAD. It would then send an X.29 message to set parameter 8 back to 0 so that the user and host can communicate once again.

| 9 | **Padding** | (Padding after carriage return.) This defines a delay that the PAD should make following the sending of a Carriage Return. |

This parameter is intended for use with slow printing devices that require a pause after Carriage Return is sent, so that the print head can return to the left margin ready for the next line of characters. The value of the parameter is the number of NUL characters that are sent.

| 10 | **Line fold** | (Line folding.) This informs the PAD of the width of the screen. |

This indicates to the PAD where it should insert newlines into the data stream, and therefore allows the PAD to take responsibility for formatting the screen. A value of zero turns the feature off.

| 11 | **Speed** | (Binary speed of start-stop mode DTE.) This is the speed of the terminal and is a read-only parameter set by the PAD. |

This parameter allows the host to enquire the speed of the terminal, perhaps to modify the amount of data to send to the screen. The encoding is as shown at the top of p. 55.

| 12 | **Flow** | (Flow control of the PAD.) This indicates whether the user can control the flow of data output by the PAD. |

| | |
|---|---|
| decimal 10 − 50 bit/s | decimal 3 − 1200 bit/s |
| decimal 5 − 75 bit/s | decimal 7 − 1800 bit/s |
| decimal 9 − 100 bit/s | decimal 11 − 75/1200 bit/s |
| decimal 0 − 110 bit/s | decimal 12 − 2400 bit/s |
| decimal 1 − 134.5 bit/s | decimal 13 − 4800 bit/s |
| decimal 6 − 150 bit/s | decimal 14 − 9600 bit/s |
| decimal 8 − 200 bit/s | decimal 15 − 19 200 bit/s |
| decimal 2 − 300 bit/s | decimal 16 − 48 000 bit/s |
| decimal 4 − 600 bit/s | decimal 17 − 56 000 bit/s |
| | decimal 18 − 64 000 bit/s |

If this is set to 1 then the user can stop data being sent from the PAD to the screen by typing XON and XOFF characters at the keyboard. This is done by control shift S (XOFF) and control shift Q (XON). This may need to be done if the host has sent a large volume of output covering more than one screen. Note that there is no parameter in the 1980 X.3 akin to parameter 10 that tells the PAD how many lines to send before waiting for a "carry-on" signal, However, the 1984 X.3 contains a PAGE parameter for this function. A value of zero means that XON and XOFF characters are buffered as every other character and sent to the host.

| 13 | Linefeed | (Linefeed insertion after carriage return.) This defines whether the PAD should insert a linefeed character after a Carriage Return either from the user or from the host, or for data echoed locally by the PAD. |
|---|---|---|

This again reflects the responsibility of the PAD to format the data appearing on the screen. If the parameter is zero then no action is taken, otherwise it is bit-oriented as for parameter 3 with the following meanings:

| | |
|---|---|
| bit 0 (decimal 1) | Instructs the PAD to insert a linefeed after each Carriage Return sent from the host to the screen. |
| bit 1 (decimal 2) | Instructs the PAD to insert a linefeed after each Carriage Return sent from the keyboard to the host. |
| bit 2 (decimal 4) | Instructs the PAD to insert a linefeed after Carriage Returns echoed from the keyboard to the screen. |

This parameter is only effective in data transfer state.

| 14 | LF Pad | (Padding after linefeed.) |
| | | This defines a delay that the PAD should make following the sending of a linefeed. |

This parameter is similar to parameter 9, except that it inserts a delay after linefeed characters sent to printing devices.

| 15 | Edit | (Editing.) |
| | | Allows local editing of data in the PAD buffer by the characters defined in the next three parameters. |

Data stored in the PAD buffer waiting for a forward condition may be edited by the user if this parameter is set to 1. If it is zero then the editing characters defined in the following three parameters have no local effect and are buffered for transmission to the host. If the parameter is set, then parameter 4 (timeout) is disabled even though it may be set. This is necessary because there would otherwise be confusion over characters already forwarded due to the timer and subsequently deleted from the buffer.

| 16 | Ch del | (Character delete.) |
| | | Defines the character deletion character. |

If parameter 15 is set to 1 then typing this character has the effect of deleting the previous character from the PAD buffer. The PAD will normally display this on the screen with a backslash (\) followed by the deleted character. It is possible, though not standard, for the PAD to determine whether a screen device is being used when the port is configured, and to actually erase the character from the screen by suitable cursor movements such as backspace-space-backspace.

| 17 | Buf del | (Line delete.) |
| | | Defines the buffer deletion character. |

If parameter 15 is set to 1 then typing this character has the effect of deleting the entire contents of the PAD buffer. The PAD will display this on the screen by showing three upper case Xs (XXX), though again it may be able to erase the characters from the screen.

| 18 | Buf echo | (Line display.) |
| | | Defines the buffer display character. |

If parameter 15 is set to 1 then typing this character causes the PAD to display the current contents of the buffer. This would be needed on a standard PAD if several character deletes had been performed and the display was consequently confusing.

## 2.10  X.28 commands and service signals

This section shows a list of X.28 commands that the user may give, and service signals that the PAD may report.

The user activates the terminal line in a manner defined by the PAD manufacturer. This will normally be by pressing the Carriage Return key, but may also be by raising signals on the interface cable. When Carriage Return is used, several may be needed if the PAD has to detect and synchronize the speed of the terminal.

The PAD will then respond with a message again defined by the manufacturer, but usually giving the identity of the PAD and some announcement or help text. The PAD prompt will then be displayed, indicating readiness to receive a command. This is *PAD Command State*.

The user would then make a call to the selected host service, would receive the COM (call connected) message, and would be in conversation with the host. This is *Data Transfer State* and the function of the PAD is defined by the X.3 parameter settings. Essentially all data is transferred between the user and the host. Any data typed by the user − even if it is a PAD command − is sent to the host; the PAD will not intercept it.

The user can escape from Data Transfer State either by using the escape character defined in parameter 1 or by using the Break key if this is defined in parameter 7. The escape character is normally DLE (control shift P). The user is then in PAD Command State and a PAD prompt will be displayed. The user can give any PAD command to control the call and the call will remain active in the background. Having given the commands, the user can return to Data Transfer State and continue the host session either by typing DLE or by using Carriage Return.

Note that calls are also possible from the network into the terminal. So for instance the terminal may burst into life and say "a call has been received from a host service at network address 754347". The user can control such a call in the same way as if the user had initiated it.

### 2.10.1  User commands

**PAR?**     Displays the values of the X.3 parameters on the screen. The command can be modified to PAR?1, 3, 6 for example to show only selected parameters. The PAD response is to return pairs of parameter numbers and their current settings. The response to the request shown above might be PAR1:1,3:2,6:1.

**SET?**     Sets the value of one or more of the X.3 parameters. SET will simply set parameters; SET? sets the parameters and requests confirmation of the setting. The confirmation will be reported by the PAD in the same way as shown for PAR command. If any of the parameters is invalid — it may be read-only or perhaps a number greater than 18 — then the confirmation message is sent to show the error. The command format is pairs of parameter numbers and required values, for example to set parameter 1 to 1 and parameter 3 to 4 the command would be SET?1:1,3:4.

**PROF**     Sets all of the X.3 parameters to a set of default values. There are several sets of these default values, and the command names which one is required. Thus PROF A2 sets the parameters to those shown in profile list A2. The profiles are chosen to be useful sets of parameters.

**CLR**     Instructs the PAD to clear the call down.
The response will be CLR CONF (clear confirmed).

**STAT**     This allows the user to enquire whether a call is currently in progress or not.
The response is either ENGAGED or FREE. Many manufacturers add to this response with information such as how long the call has been in progress and how many packets have been sent.

**INT**     Instructs the PAD to send an Interrupt packet.
This will normally stop the host computer from what it is currently doing.

**RESET**     Instructs the PAD to send a Reset packet to reinitialize the Virtual Circuit.
This is covered in more detail in Chapter 4.

**xxxx**     Makes a call to the given network address. Many PADs also provide a more explicit command such as CA xxxx.
The command is made up of three parts: facility requests, the called address, and *Call User Data*. These terms are all covered in Chapter 4.
The format of the facility request is
    N followed by the NUI of the user.
    R to request reverse charging.
    G followed by a closed user group number.
Any number of these three may be specified and they are separated by a comma (,). If any are specified then they are followed by a hyphen (-).
The format of Call User Data is
    D followed by the data — this will be echoed to the terminal.
    P followed by the data — this will not be echoed.
The normal type of call command would be simply 76543778. A command requiring all features might be

    N85483743,R,G03-76543778DNORMAL

## 2.10.2 Service signals

**PAR**  Used to indicate the values of X.3 parameters in response to the SET, SET?, and PAR commands.

It is followed by number/value pairs, for example PAR1:1,3:2.

**CLR**  This is used to indicate that the call has been cleared down.

It is followed by a reason as shown below. All but one of these is due to the call being cleared remote from the user, either by the host service or by the network itself.

| | |
|---|---|
| CLR CONF | user cleared the call |
| CLR OCC | called number engaged |
| CLR NC | network congestion |
| CLR INV | invalid facility requested |
| CLR NA | access to host barred |
| CLR ERR | local procedure error |
| CLR RPE | remote procedure error |
| CLR NP | called number not assigned |
| CLR DER | called number out of order |
| CLR PAD | PAD cleared after X.29 invitation |
| CLR DTE | host service cleared call |
| CLR NRC | host does not subscribe to reverse charging |

**COM**  Indication that the call is connected.

This signal is also sent following a successful incoming call to the terminal. In this case it is sent with a message announcing the call, and a message giving the address of the entity making the call.

**XXX**  The line deleted signal.

This is explained under editing with X.3 parameter 15.

**RESET**  Incoming Reset.

Resets are explained in more detail in Chapter 4. This signal indicates that an incoming Reset has been received and will be qualified as follows:

| | |
|---|---|
| RESET DTE | host performed Reset |
| RESET ERR | there was a local procedure error |
| RESET NC | the network Reset owing to congestion |

**Break**  Indication of Break.

The PAD sends a Break condition to the terminal if the host sends an X.29 Indication of Break message.

**ERROR**             Error in command.

This is given if the user enters an invalid command in PAD Command State.

**ENGAGED/FREE**     These are responses to the STAT command.

### 2.10.3  X.28 standard profiles

Two standard profiles are defined in X.28 as follows:

*Transparent standard profile*

> 1: 0   2: 0   3: 0   4: 20   5: 0   6: 0   7: 2   8: 0   9: 0
> 10: 0   12: 0   13: 0   14: 0   15: 0   17: 24

and 11 set to the terminal speed. This makes the PAD as transparent as possible to data from the user.

*Simple standard profile*

> 1: 1   2: 1   3: 126   4: 0   5: 1   6: 1   7: 2   8: 0   9: 0
> 10: 0   12: 1   13: 0   14: 0   15: 0   17: 24

and 11 set to the terminal speed. This corresponds to an everyday default setting that will allow initial communication with many host services.

## 2.11  Differences in 1984 recommendations

This section shows the major additions made to triple-X in the 1984 recommendations.

### 2.11.1  X.3 (1984)

The major addition to X.3 is four new parameters. These have the following functions:

| Number | Name | Description |
|---|---|---|
| 19 | **Display** | (Editing PAD service signals.) This parameter indicates whether the terminal attached to the PAD is a printing type or a screen type, and thus determines what happens when the editing characters are used. |

The values of the parameter are as follows:

| | |
|---|---|
| decimal 0 | No editing PAD service signals |
| decimal 1 | Use editing PAD service signals for printing device |
| decimal 2 | Use editing PAD service signals for display device |
| decimal 8 and 32–126 | Use one character for editing PAD service signals |

When parameter 6 is 1 (enabling service signals) then the value of parameter 19 determines the editing procedures that come into effect when the characters defined in parameters 16, 17, and 18 are typed by the user. As long as parameter 15 is set to 1 then the edit takes effect on the buffer as described for the 1980 version. What the new parameter does is to define the effect on the screen or paper that the user sees.

If it is set to 0 then no service signals are displayed so the deleted text remains displayed. Other values of the parameter have the effect shown below:

| Parameter value | Effect for character delete | Effect for line delete |
|---|---|---|
| 1 | Backslash (\) displayed. | XXX displayed. |
| 2 | Character 'rubbed out' with backspace, space, backspace sequence. | Line 'rubbed out' with backspace, space, back- space sequences. |
| 8, 32–126 | Character from IA5 dis- played. | XXX displayed. |

This parameter standardizes the practices that were common in PADs for deletion of buffer contents.

**20    Echo mask**    (Echo mask.)
This parameter is additional to parameter 2 (echo) and defines a set of characters which are not echoed back to the terminal.

If parameter 2 is set to 1 then characters received from the keyboard are echoed back to the display. Parameter 20 stops the echoing of certain characters as follows:

| Decimal value | Characters not echoed |
|---|---|
| 0 | None (all characters echoed) |
| 1 | Carriage return (CR) |
| 2 | Linefeed (LF) |
| 4 | VT, HT, FF |
| 8 | BEL, BS |
| 16 | ESC, ENQ |
| 32 | ACK, NAK, STX, SOH, EOT, ETB, ETX |
| 64 | Editing chars defined by params 16, 17, 18 |
| 128 | DEL and all other chars not mentioned above in columns 0 and 1 of IA5 |

Additionally, the following are not echoed: XON and XOFF if parameters 5, 12 or 22 are non-zero; the PAD recall parameter defined in parameter 1.

Any combination may be selected by adding the appropriate values.

**21    Parity**    (Parity treatment.)
Defines the parity checking and generation performed by the PAD.

A value of zero indicates no parity generation or checking. A value of 1 indicates that the eighth bit of characters received from the keyboard is checked for correct parity. A value of 2 indicates that the eighth bit of characters sent to the display has the parity bit set, both for host data and PAD service signals. A value of 3 sets both parity checking and generation. If an error is detected in an incoming character then it is discarded and not echoed, and a service signal is displayed.

**22    Page wait**    (Page wait.)
Allows the PAD to scroll data to the display.

If the parameter is set to 0 then page wait is disabled. If it is set to a value between 1 and 255 then it defines the number of lines to send per scroll. A line is defined as a sequence of characters terminated by a line feed.

## 2.11.2   X.29 (1984)

The major addition to X.29 is the Reselection message. This is sent from the host to the PAD and invites the PAD to clear the call and then make a new call to another host defined in the X.29 message. This is a type 07 message, and the information to make the new call is encoded in bytes 2 onwards.

## 2.11.3   X.28 (1984)

The major additions to X.28 are to support the extra four X.3 parameters. New service signals are defined as follows:

- \ and the backspace-space-backspace sequences for the editing commands.
- An undefined signal indicating that an error has been detected in the parity of an incoming character.
- The signal PAGE which indicates that the number of lines defined by parameter 22 has been displayed.
- The signal TRANSFER which indicates that an X.29 Reselection message has been received, and a new host service is being called.
- The PAD prompt is now defined as *.

Once the page wait condition has been detected and the service signal displayed, then no more data will be displayed until the condition is cleared. This is done by one of the following:

- The user typing XON, the page wait cancellation
- Parameter 22 being set to 0
- A data forwarding condition
- The echoing of a linefeed
- After the line deleted service signal has been sent
- Leaving PAD command state.

Two profile identifiers are defined for use with the PROF command. These are PROF 90 which selects the simple standard profile, and PROF 91 which selects the transparent standard profile.

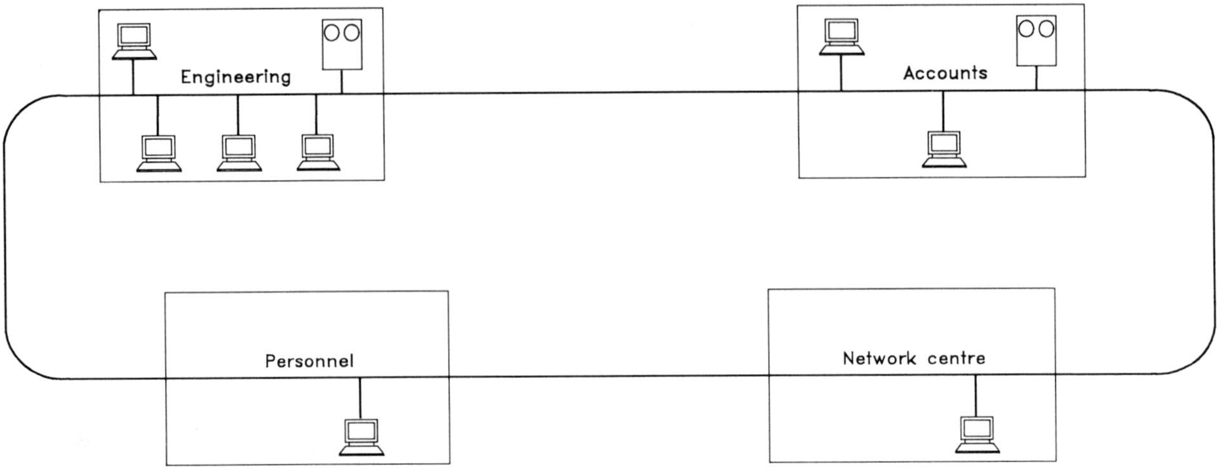

**Fig. 3.1**   Theoretical ring topology

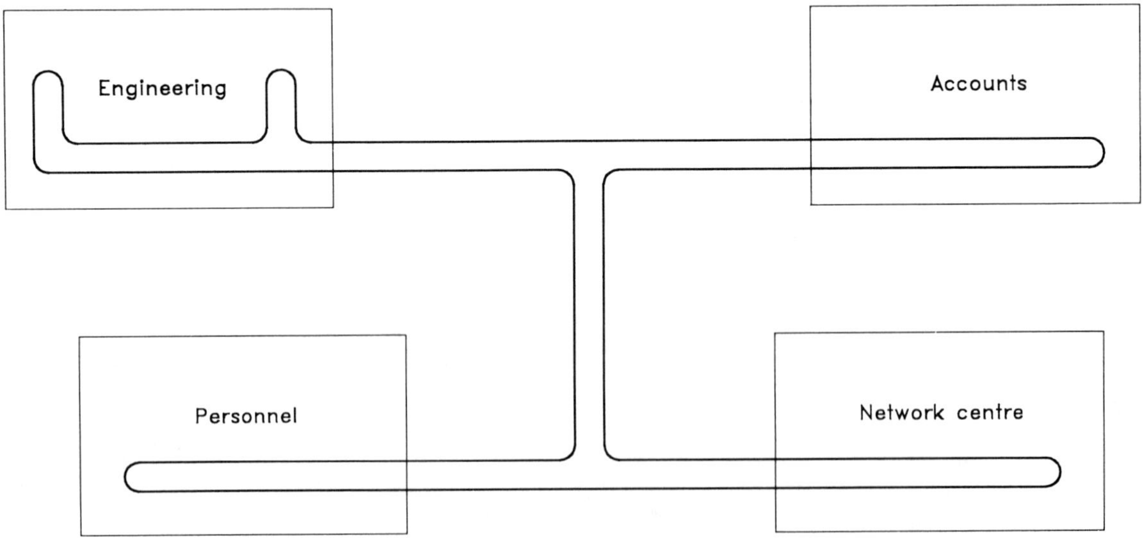

**Fig. 3.2**   Actual ring layout

# 3 Topology and components

## 3.1 Introduction

A Local Area Network (LAN) is a private network covering a radius of up to a few kilometres. It typically covers one site of an organization and links all the users and services on that site. It usually also has a connection to a larger network which may be public or private, and which links sites of the organization together and may also provide access to third party services. The larger network is referred to as the Wide Area Network (WAN). To all intents and purposes WANs have been restricted to X.25, and it is only recently that new technologies have become available.

The previous chapter showed how a LAN could be constructed using X.25 PADs, switches and combined switch/PADs and it has been shown in previous chapters how this provides a service to the site, in which errors are detected and data retransmitted automatically. On the local scale there are other technologies available which offer some advantages over X.25 mainly in the area of speed of transmission.

## 3.2 Ring topologies

This technology is divided into the *Slotted Ring* and *Token Ring* systems. As the name suggests, there is a closed loop – a ring – of cable which is routed around all points on the site that require access to the network. This is shown in Fig. 3.1 for an example site.

Of course most buildings do not lend themselves to actually having a circular ring. There are corridors, stairs, and cable ducts which dictate where the cable can lay, and the users and services will be in inconvenient places. The actual layout may be as shown in Fig. 3.2.

One thing that has to be considered with a ring technology is: what happens if there is a break? The whole network will be down until the fault can be identified and rectified. This is usually addressed by having several sectors to the ring that can be isolated from each other. If there is a fault the sectors can be isolated in turn until the faulty one is found and the other sectors can then carry on running while the fault is fixed. This is shown in Fig. 3.3.

Engineering

Accounts

Personnel

Network centre

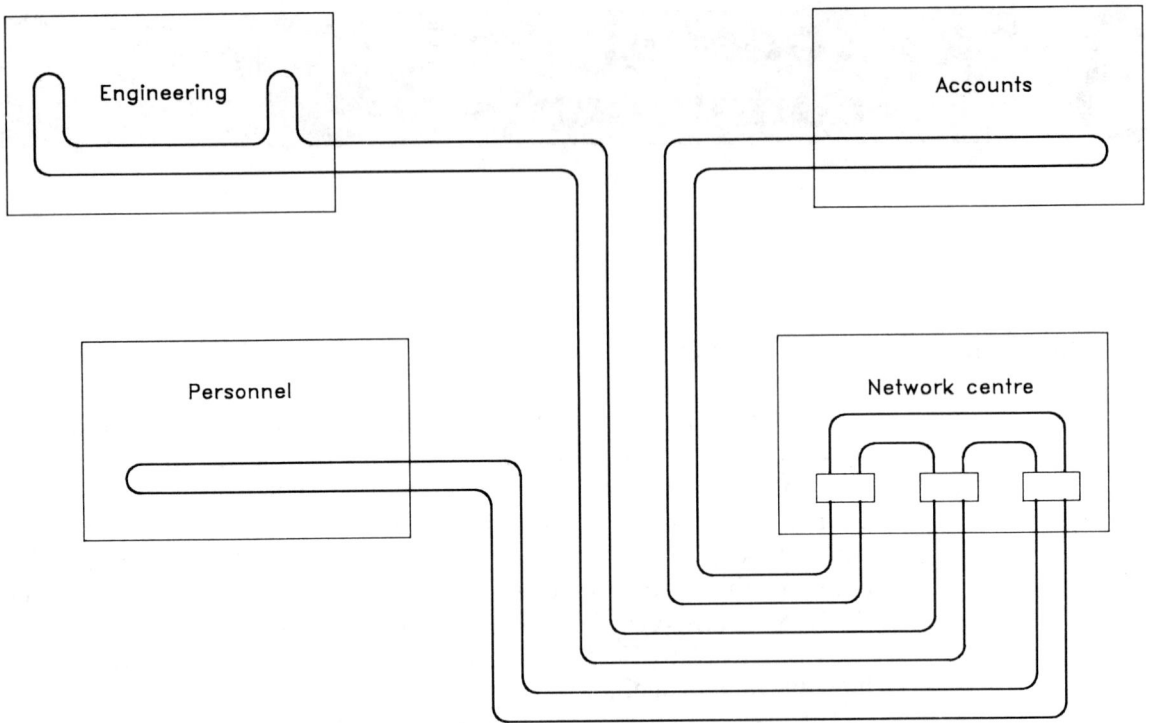

Sector included
in the ring

Sector isolated
from the ring

**Fig. 3.3** Segmenting the
ring to allow for fault
isolation

### 3.2.1 Cambridge Ring

Cambridge Ring is the best example of slotted ring technology. A fixed number of slots, each of which can carry a fixed amount of data, continually travel around the cable. A sender waits for an empty slot to come around, puts the data in, and inserts the address of the receiver and its own address. This slot then carries on around the ring. The receiver detects its address in the slot and extracts the data, and also sets the received flag. The slot circles back to the sender which then empties the slot and can see whether or not it was received.

The slot is fixed at 40 bits long and has the following structure. This is defined in the CR82 specification (see Appendix D).

First bit onto ring

| | |
|---|---|
| 1 | Slot framing bit (always 1) |
| 2 | Full/empty marker |
| 3 | Monitor pass bit |
| 4–11 | Destination address |
| 12–19 | Source address |
| 20–27 | Data |
| 28–35 | Data |
| 36–37 | Type bits |
| 38–39 | Response bits |
| 40 | Parity bit (set for even parity) |

Last bit onto ring

The slot is referred to as a *minipacket*, and has the ability to carry exactly two data bytes. The fields in the minipacket are used as follows:

*Full/empty marker*
Set to 1 when the slot is in use; set to 0 if the slot is empty and available.
*Monitor pass bit*
This bit is set to 1 by the source and reset to 0 when the minipacket passes the special monitor station on the ring. If the minipacket passes the monitor a second time then there is a fault in either the source or destination. The monitor then resets the full marker so that the minipacket may be used by others.
*Response bits*
These bits allow the destination to indicate what action has been taken on reception of the minipacket. The bits are both initialized to 1 by the source. Encodings are:

| bit 38 | bit 39 | |
|---|---|---|
| 1 | 1 | Ignored. The bits have not been changed by the receiver so it has ignored the minipacket. |
| 1 | 0 | Not Selected. Each address may be individually set to listen to minipackets coming from everyone, from no-one, or from a particular address. This response implies that the destination was not listening to minipackets from this source. |
| 0 | 1 | Accepted. The minipacket has been read. |
| 0 | 0 | Busy. The destination acknowledges the minipacket but could not process it – perhaps because it is still dealing with a previous one. |

There are only two types of component that may physically connect to the ring cable: a power supply and a repeater. The power supply simply supplies 28 V to the ring cable and powers the repeaters that are connected. There may need to be several power supplies if the ring is long and they must be physically spread around the ring. One of these is designated as the master power supply and the others are slaves. The slaves only apply power to the ring if the master is doing so, thus the master determines whether the ring is operational or not. The *repeaters* boost the signals on the cable much like modems do on a normal signal line; they also provide the only access point for other equipment to attach to the ring and exchange data. The maximum distance between repeaters depends on the type of cable used and is typically between 200 m and 500 m.

Since repeaters are the only means of connection to the ring, and since they are powered from the ring, nothing that may happen to external equipment can affect the integrity of the ring.

Every ring has exactly one *monitor* which connects to the ring via a repeater. The monitor will determine the time taken for a bit to travel around the ring cable and through all the repeaters and from this will be able to determine the number of slots that can circulate.

The monitor may contain a 40-bit shift register so that the effective length of the ring is extended. Then, no matter how short the cable is, at least one slot can circulate. If the length is not a whole number of slots then a gap full of zero bits will be inserted. Once all this has been done the ring is said to be *synchronized* and the slots and a possible gap will continue to go round like a train on a track.

The ring operates at a normal speed of 10 Mb/s but this will be modified slightly by the monitor to ensure that a whole number of bits circulates. To connect to the ring, a station is attached to the repeater. The station is powered from the external equipment with a 5 V supply, and simply takes care of gating the minipackets on to and off the ring. The combination of station and repeater is called a *node*. A typical ring is shown in Fig. 3.4.

Cambridge Rings offer much higher speeds than X.25 even though the full 10 Mb is not all available for data. The minipacket layout means that only sixteen of the forty bits can be used for data, and this capacity is further eroded by the higher-level protocols.

The main disadvantage is the cost of connection. The nodes and repeaters are implemented with discrete logic, so the cost of a node and PAD is very high compared to that of an X.25 PAD. The topology means also that there is no concept akin to the switchpad. Widely separated groups of users all require ring cable, node, and PAD combinations.

Despite some early promise, host implementations of Cambridge Ring are few and far between, so in general hosts have to connect either through Reverse PADs with all the disadvantages described in Chaper 2, or via X.25 using gateways.

### 3.2.2 Token ring

Token ring systems are largely similar to the empty slot system, but there is no fixed slot size. In the idle state, when none of the stations are

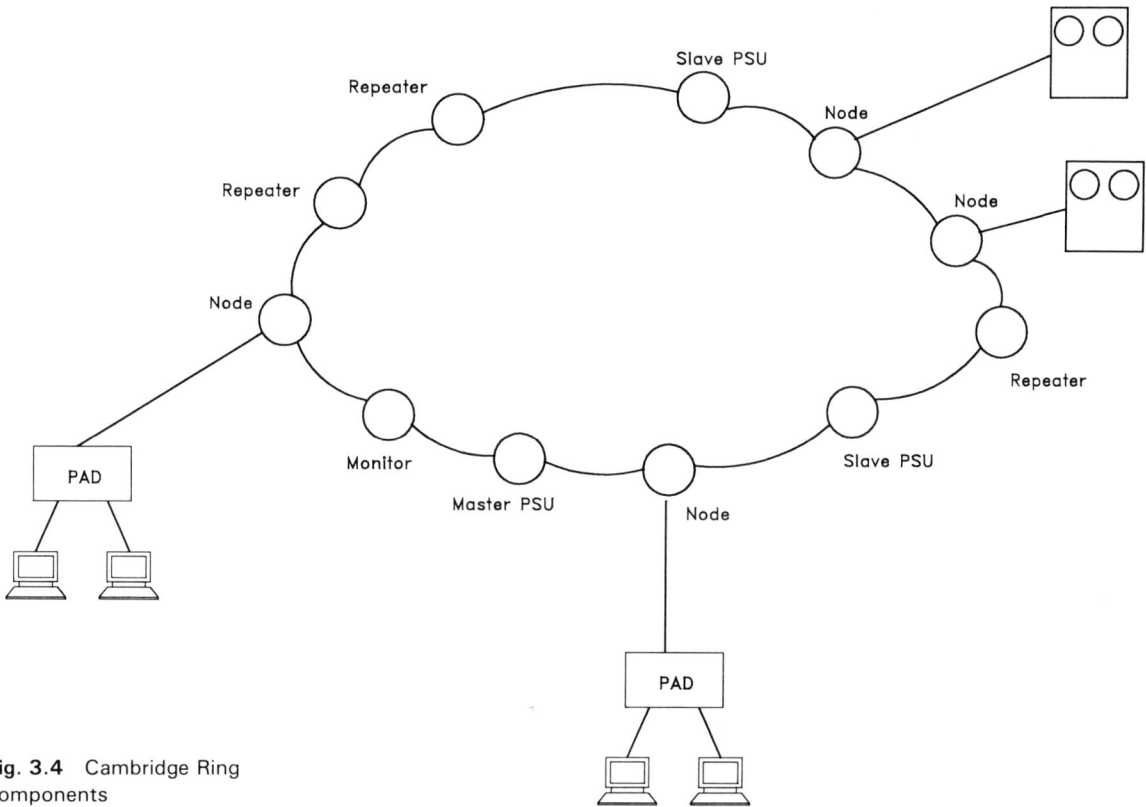

**Fig. 3.4** Cambridge Ring components

using the ring, a simple structure called the token circulates around the ring indicating the idle condition. When a station wishes to send data then it waits for the token, marks the token as busy, and inserts data following the token. The data includes source and destination addresses as for the Cambridge Ring system, and similarly the whole slot cycles back to the sender. When the sender receives the slot back then it must release the slot by marking the token as free and allow it to circulate to the next station. If the length of ring allows then it is possible for several tokens to circulate and for several data slots to be available.

An adaptation of the token ring system allows a station to use a priority over others in gaining the token. Each station has a preassigned fixed priority. If the station has data to send and the token is in use, then the station examines the priority field in the slot. If it is less than that of the station, then the station stores the lower priority internally and inserts its own priority in the slot. Assuming the priority is not usurped by a still higher one, then the slot will arrive empty at the station on its next circle and can then be used. When the station has finished with the slot then the token is freed, and the lower priority previously stored in the station is returned to the priority field of the slot.

IBM token ring is an example of a system that uses the priority additions.

## 3.3  Bus topologies

In a bus system a single cable links all points of the site as in a ring system, but the ends of the cable are not joined together. Fig. 3.5 shows how a bus might address the need of the network shown in Figs. 3.1 and 3.2.

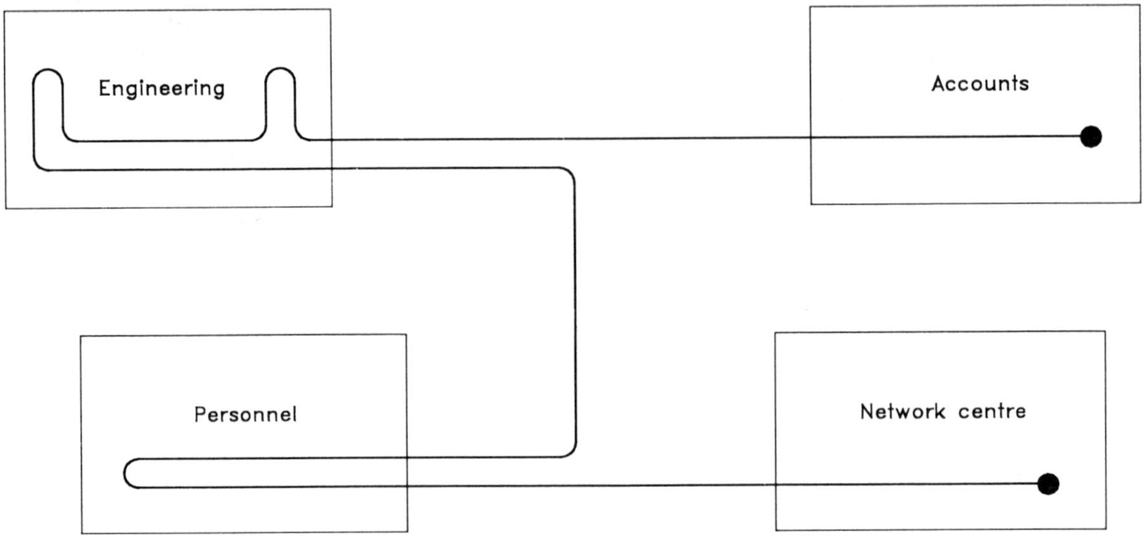

Fig. 3.5  Layout of bus system

The network is still subject to the physical restrictions imposed by the building, and in terms of topology there is little practical advantage of bus systems over ring systems.

### 3.3.1  Ethernet

For many years Ethernet has been the only bus topology LAN that is commercially practical. It was developed from work done on the ALOHA project at the University of Hawaii, and the first commercial work was done by the Xerox Corporation and the Digital Equipment Corporation. Whilst new bus systems are becoming available, Ethernet is still being developed and is likely to be around for some time. The components of the network are shown in Fig. 3.6.

The cable is traditionally a thick coaxial type that is defined to be yellow. The cable must be handled with some care to avoid kinks and other physical damage, and much be laid so that the bends have a radius of not less than 250 mm. The cable is marked every $2\frac{1}{2}$ metres, and the ends and any connections to the cable must be made at those points. All of this is to prevent internal reflections of the signal and consequent data corruption. The ends of the cable are terminated by a resistor which matches the impedance of the cable and again reduces reflections. More modern systems may use thin coaxial cable or even twisted-pair wiring with

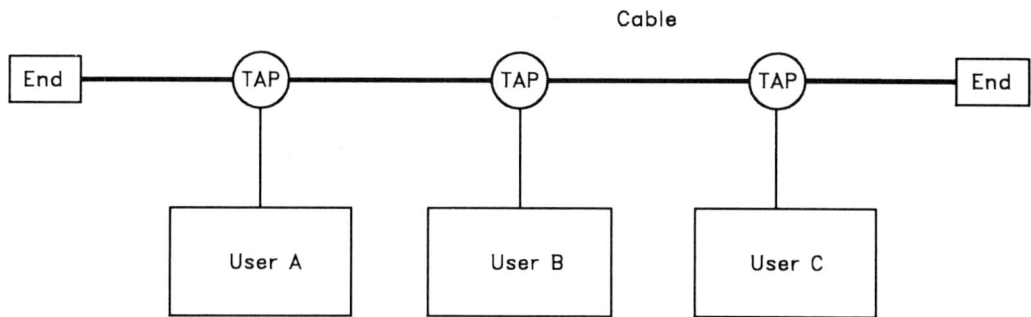

Cable

End — TAP — TAP — TAP — End

User A | User B | User C

consequent reductions in cost and difficulty of installation. Connections to the cable are made by devices called TAPs which are powered from the user device and which transfer data between the user and the cable.

There is no implied data structure of information on the cable. A user, via its TAP, can transmit data onto the cable, and this data can be detected by all other users attached to the cable. A protocol is imposed upon all users in order to ensure that reasonable data transfer — that is, organized and correct — may be carried out. This protocol is called *Carrier Sense Multiple Access with Collision Detect*, and is designated CSMA/CD.

- When a user is sending data then it does this by modulating a carrier signal, and other units can therefore detect that the cable is in use by sensing the presence of a carrier.
- When a user wishes to send then it waits until any carrier signal from other units ceases, and then sends its own data on its own carrier.

The mechanism allows all units to send on the cable, one at a time, by waiting for each other to finish. There is however a problem because if two units want to send then they will wait for the carrier to cease, and then both start transmitting together. The rule shown above is therefore modified:

- When a user wishes to send then it waits until any carrier signal from other units ceases, and then sends its own data on its own carrier. As it does so it "listens" to what is on the cable, which should obviously be its own data. If the data is not the same as what is being transmitted, then two or more units are transmitting at the same time. If this happens then the user stops transmitting and waits a random time before trying again.

The situation of two or more users transmitting at the same time is referred to as a *collision*. The collision detect mechanism ensures that all participants in the collision "back off", and by doing so for a random time there is little chance of a future collision between the same participants.

Collision detection is a little more complex than simply listening to the first few bits of data, because the signal takes a finite time to travel along the cable. Fig. 3.7 shows a situation in which station A starts to transmit

to station C which is very close. The signals transmitted by A will take a little time to reach station B, and in that time B may also start to transmit. Both A and B have therefore obeyed the CSMA/CD procedure, but unless they continue to listen for some time there may still be a collision after the start of the transmission. The time in which such a collision is possible is dependent on the distance from A to B, and therefore on the length allowed for the Ethernet cable.

The Ethernet specification defines the maximum cable length to be 500 m. The transmitting station therefore has a definite window during which a collision is possible. A station that detects a collision will force a special condition on the cable called a Jam, as an indication that corruption has occurred and that back-off action is required. The Jam signal is a special bit sequence that the standard defines as being between 32 and 48 bits long.

When a user gets control of the cable then there is no limit imposed by the CSMA/CD mechanism on how much data can be sent, or what structure the data should have. That is a problem for the higher layers of protocol to deal with. There may also be limitations imposed by the particular hardware used.

## 3.4  Protocols

The situation regarding protocols on Local Area Networks has become much more organized than it was in the early nineteen-eighties, and there is now a coherent set of standards common to the various technologies. These standards also allow for new LAN technologies to be introduced without upsetting the standardization already done.

Going back to Chapter 1, the key to X.25 is the use of the lower three layers of the ISO OSI seven-layer model. These same three layers are used in LANs. For use in the LAN world, the data link layer (layer two) has been further subdivided in order to cater for the different technologies that exist and that will be developed. This rationalization work was carried out by the Institute of Electrical and Electronic Engineers (IEEE) in the USA, and the result is a series of standards:

| | |
|---|---|
| 802.1 | Higher Layers and Management |
| 802.2 | Logical Link Control |
| 802.3 | CSMA/CD |
| 802.4 | Token Bus |
| 802.5 | Token Ring |
| 802.6 | Metropolitan Area Networks |
| 802.7 | Slotted Ring |

**Fig. 3.8** Layout of LAN protocol layers

| ISO OSI layers | X.25 | |
|---|---|---|

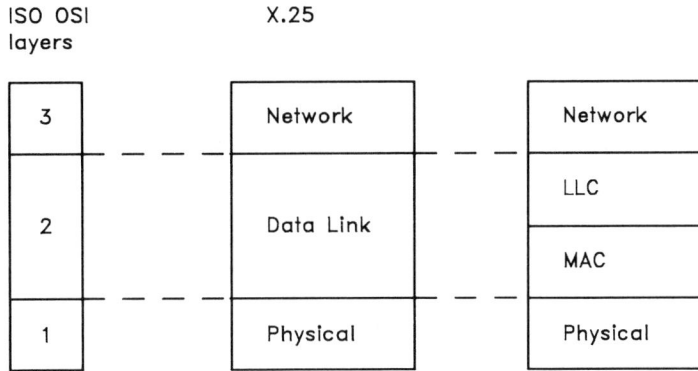

These standards are being accepted by ISO as the 8802 series, and given a little more time they will be as solid as X.25. Fig. 3.8 shows how the protocols are arranged.

The data link layer has been split into two sub-layers: *Media Access Control* (MAC) takes care of the specific network technologies, and *Logical Link Control* (LLC) provides the mechanism for error detection and checking required by layer three. Thus, Ethernet networks have a MAC layer defined in 802.3 to provide CSMA/CD and token ring networks have a MAC layer defined in 802.5.

The interface beteen layers is called a *Service Access Point* (SAP) — because one layer provides a service for the one above it — and the diagram now changes to that shown in Fig. 3.9. The MAC Service Access Point (MSAP) provides a means of transmitting data between users on the network, so in the case of Ethernet it uses CSMA/CD to

**Fig. 3.9** Positions of Service Access Points

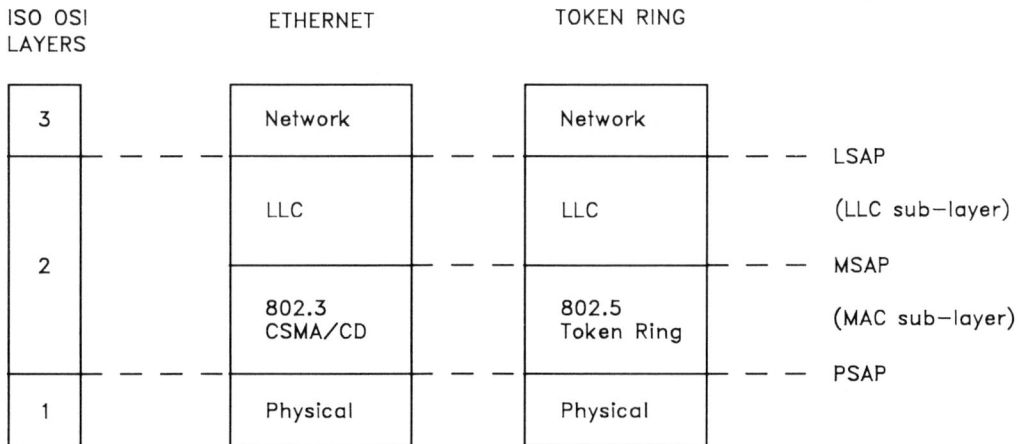

LSAP — LLC Service Access Point

MSAP — MAC Service Access Point

PSAP — Physical Service Access Point

73

provide a service to LLC. The MAC protocol provides error detection by the use of a Frame Check Sequence as discussed in Chapter 1.

The LLC layer may provide correction of errors by using methods similar to X.25 layer two, such as the N(R) and N(S) values, and REJ frames. In this case the combination of MAC and LLC therefore provides a layer two service via the LLC Service Access Point (LSAP) that is equivalent to X.25 layer two. Such a service is called LLC type two, and is said to be connection-oriented. Such a service gives a "pipe" for the data, which can then be passed to the LSAP by layer three without any addressing information because there is a virtual circuit.

Alternatively, LLC may conform to LLC type one, which is a connectionless service. Here there is no virtual circuit so every unit of data that is transmitted has to include its source and destination address so that it can be routed through the network. In a connectionless system the units of data sent between two components may go via different routes at different speeds and arrive out of sequence. They may also be lost in the network with no indication to either end.

LLC is said to be class one (Class I) if it implements type one only, or to be class two (Class II) if it implements both type one and type two.

One option for manufacturers is to run X.25 layer three over a Class II LLC on a LAN. This is defined in the ISO 8878 standard, Use of X.25 to provide connection mode service. X.25 is simply a protocol, and there is a further standard, ISO 8880/2, which defines how the service is made available. It defines how the layer three functionality is interfaced in the Network Service Access Point (NSAP). This is called the *Connection Oriented Network Service* (CONS). X.25 layer three cannot run over a Class I LLC so a different higher layer has to be chosen, usually TP(4). The service is then made available via the NSAP and is defined in ISO 8880/3.

**Fig. 3.10** LAN protocol stacks

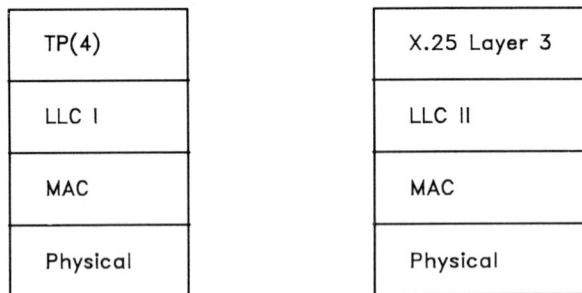

| TP(4) |
| --- |
| LLC I |
| MAC |
| Physical |

| X.25 Layer 3 |
| --- |
| LLC II |
| MAC |
| Physical |

These different protocol stacks are shown in Fig. 3.10. The differences between types of network services are resolved in layer four. This is discussed in Chapter 6.

## 3.5 Modems and line drivers

Chapter 7 shows that whatever electrical system is used at layer one to link two components together, there is a limit to how far apart the components may be. In order to extend this distance some extra components are needed – the modem or line driver. These extend the distance by modifying the electrical characteristics of the signal, and the amount of extension depends on the type of modification performed. Fig. 3.11 illustrates the basic principle.

**Fig. 3.11** Use of a modem to increase possible distance between components

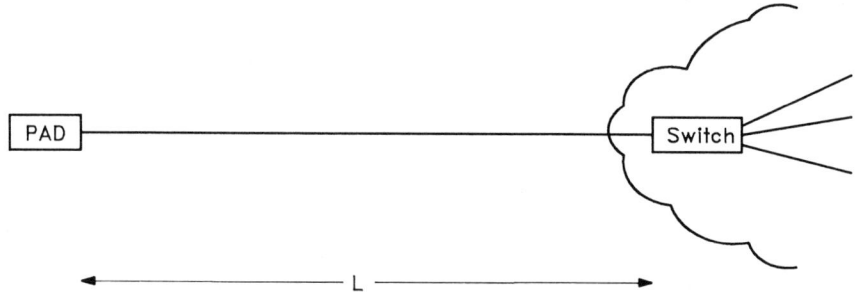

Without modems, the distance between the two components is limited to L by the electrical characteristics.

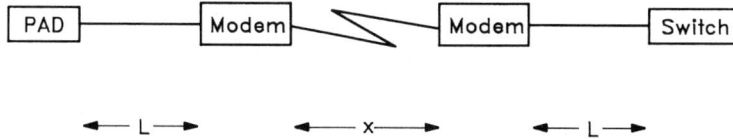

With modems, the distance between the two components is extended. The distance x depends on the type of modems used.

The simplest example of this type of component is the line driver. This essentially just amplifies the signals so that they no longer conform to the relevant standards but can travel further. Line drivers would normally only be used within a building, and typically can span a distance of a few hundred feet.

A component that is similar to the line driver is the baseband modem. This performs some encoding of the signal so that it can travel further than with a line driver, but it is still digital so cannot normally be used on the telephone network.

To span further distances requires a true modem. The modems establish a carrier signal between them which is modified, or modulated,

Signal to be
transmitted

1  0  1 1  0 0  1

Modulated
carrier signal

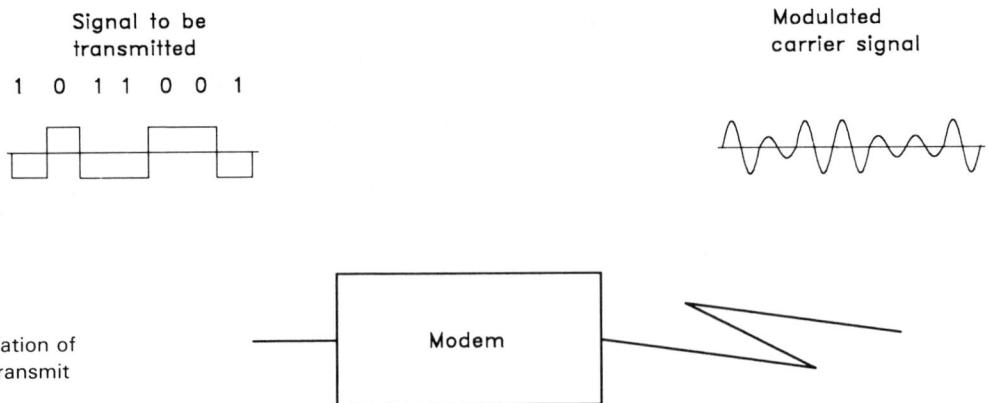

**Fig. 3.12** Modulation of carrier signal to transmit data

by the data signal to be transmitted. This is shown in Fig. 3.12. The carrier signal is usually a sine wave which is modulated to two different conditions depending on whether a 0 or 1 is to be transmitted. At the far end modem the modulated carrier is demodulated to restore the original signal. The MODulator and DEModulator give the MODEM its name.

There is also a third "no signal" level of modulation of the carrier signal. This is provided so that there is always a carrier present at some level of modulation whatever is happening, so the modems can detect that the signal path is still available. This is reported on front panel LEDs and via the interface to the equipment as Carrier Detect, so if the signal path breaks, this can be detected by the equipment and acted upon.

There are three common methods of modulation illustrated in Fig. 3.13:

● *Amplitude Modulation* is not in common usage because the sudden changes in signal level on the cable can affect adjacent equipment. Also, the signal level on the cable will be reduced by the electrical characteristics of the cable, making reliable level detection difficult.
● *Frequency Modulation* is the most popular method for low-speed modems due to its relative cheapness and reliability.
● *Phase Encoding* is used with higher-speed modems. It is somewhat complex and therefore expensive to implement, but is the method most suitable for fast and reliable communication.

In the basic encoding, a part of the sine wave is "missed out" to signal to the remote end. The conditioning of the carrier is called a phase shift and can be used to signal binary data. By causing differing amounts of phase shift, more than the two binary conditions can be communicated, and it is common for two or more data bits to be transmitted on each phase change.

In very-high-speed modems, phase shifting is combined with some amplitude modulation to signal four or more data bits with each change in modulation.

**Fig. 3.13** Different
modulation principles

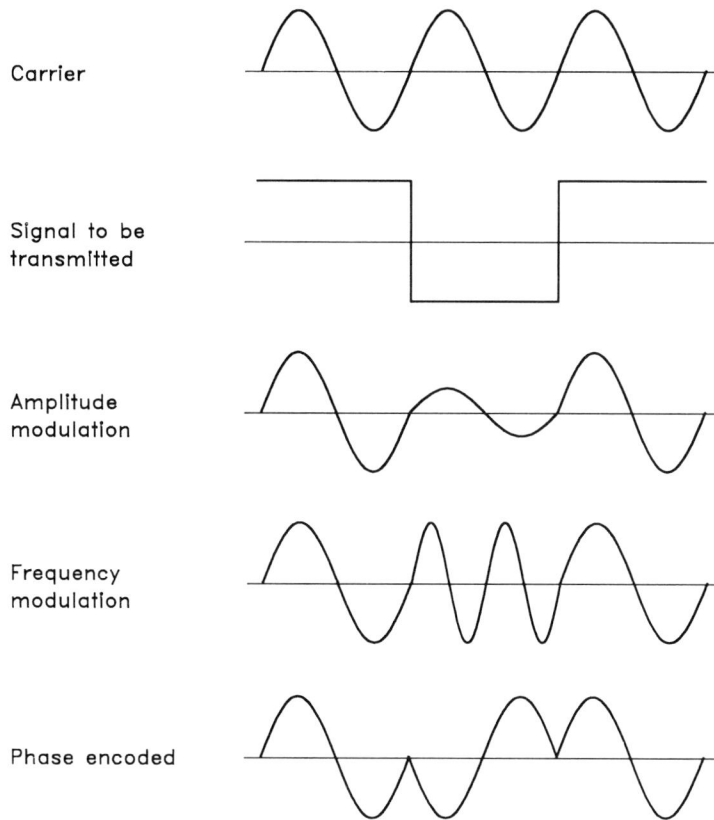

Carrier

Signal to be
transmitted

Amplitude
modulation

Frequency
modulation

Phase encoded

The reason that modulated carriers can travel further than the digital signals that they represent, is that being "rounded" they travel along the cable with less impedance. A detailed discussion of the propagation is outside the scope of this book. However, it is possible to see that the modulated signals resemble speech, and thus travel long distances in the way that speech does. The digital signal on the other hand has a very "sharp" look, and as the distances increase the signal becomes distorted and loses its sharpness. It cannot then be reliably detected at the far end.

### 3.5.1  Loopback testing

Modems may optionally contain loopback features allowing for both manual and automatic testing. There are two common loopback positions as shown in Fig. 3.14.

Local loopback state is entered by pressing a button on the modem, and has the effect of connecting the transmit and receive lines at the "back" of the modem, that is the side away from the component. Any data transmitted by the component then gets reflected back to the component and can be checked for correctness. This therefore allows the component to test itself, the cable to the modem, and the modem.

```
┌─────────────┐   ┌─────────┐                          ┌─────────┐   ┌─────────────┐
│  Component  │───│  Modem  │───────⌇─────────────────│  Modem  │───│  Component  │
└─────────────┘   └─────────┘                          └─────────┘   └─────────────┘
```

NORMAL SITUATION LINKING COMPONENTS TOGETHER

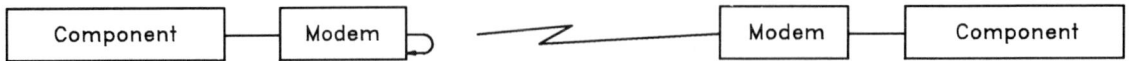

```
┌─────────────┐   ┌─────────┐                          ┌─────────┐   ┌─────────────┐
│  Component  │───│  Modem  ↺         ───⌇─────────────│  Modem  │───│  Component  │
└─────────────┘   └─────────┘                          └─────────┘   └─────────────┘
```

LOCAL ANALOG LOOPBACK

```
┌─────────────┐   ┌─────────┐                          ┌─────────┐   ┌─────────────┐
│  Component  │───│  Modem  │──────⌇──────────────────│  Modem  ↺─  │  Component  │
└─────────────┘   └─────────┘                          └─────────┘   └─────────────┘
```

REMOTE DIGITAL LOOPBACK

**Fig. 3.14** Loopback positions for modem testing

Both of the modems can be checked in this way – simultaneously if required – and therefore test the whole link with the exception of the cable between the modems.

Remote digital loopback state is entered by pressing a button on the modem and has the effect of connecting the transmit and receive lines at the "front" of the modem, that is remote from the one on which the button was pressed. This is achieved by the modems exchanging control signals. Any data transmitted by the local component then gets reflected back and can be checked for correctness. This therefore allows the component to test itself, the cable to the local modem, the local modem, the cable to the remote modem and the remote modem. Remote loopback can only be done in one direction at a time.

Using loopback facilities the whole of the transmission path between two components, including the components themselves, can be tested, and a fault isolated. This requires test software to be present in the components to transmit, receive and check data, and report on the result.

The definition of where the loopbacks are applied is in the X.150 standard, and three loopbacks are in common use:

- Local analog – designated Loop 3.
- Remote digital – designated Loop 2.
- Loopback on the actual component itself allowing tighter definition of fault reported by Loop 3. This is designated Loop 1.

Modems that implement loopbacks generally allow the loopback states

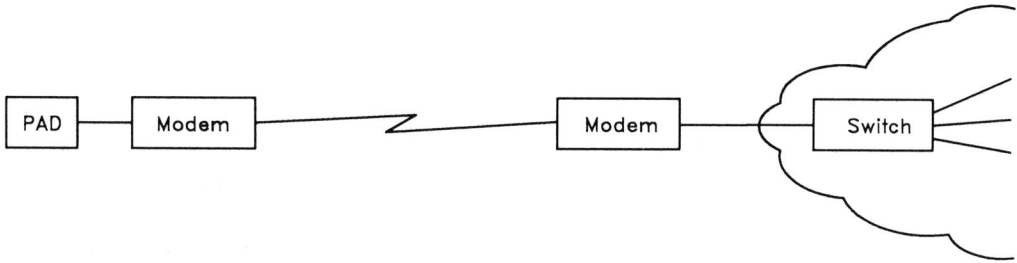

CIRCUIT IN NORMAL STATE

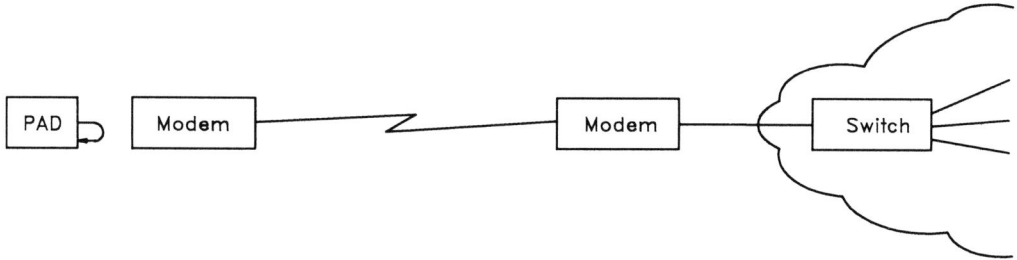

EFFECTIVE CIRCUIT AFTER 'LOOP 1' COMMAND

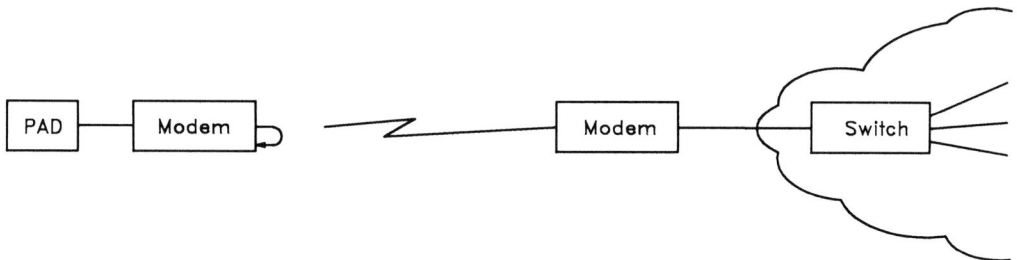

EFFECTIVE CIRCUIT AFTER 'LOOP 3' COMMAND

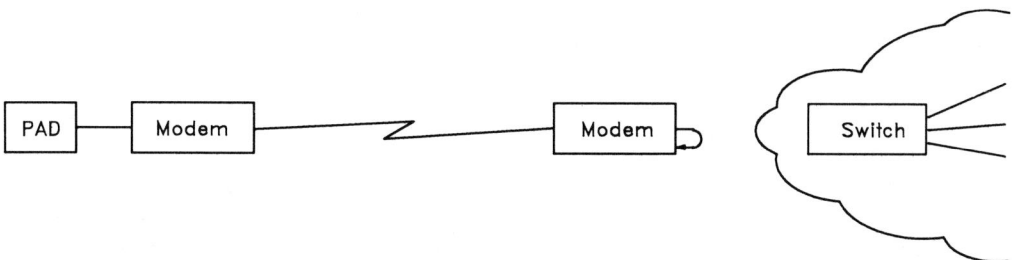

EFFECTIVE CIRCUIT AFTER 'LOOP 2' COMMAND

**Fig. 3.15** Effective circuits with loopbacks applied

to be entered automatically by the components sending control signals across the interface. The most common form of this is defined in the V.54 standard, and this is covered in more detail in Chapter 7. Because of this, modems that perform loopback are often referred to as V.54 modems.

The use of V.54 means that the entire testing process can be carried out by commands given to the component. Thus a PAD connected to a switch as shown in Fig. 3.15 can test out the complete circuit up to and including the modem on the switch, by commands given at a local terminal.

If all of these checks work satisfactorily then the rest of the circuit could be checked using software in the switch and a command terminal connected to it. See Fig. 3.16. The types of command used to perform testing are illustrated in Fig. 3.17.

**Fig. 3.16** Effective circuits after loopback from switch end

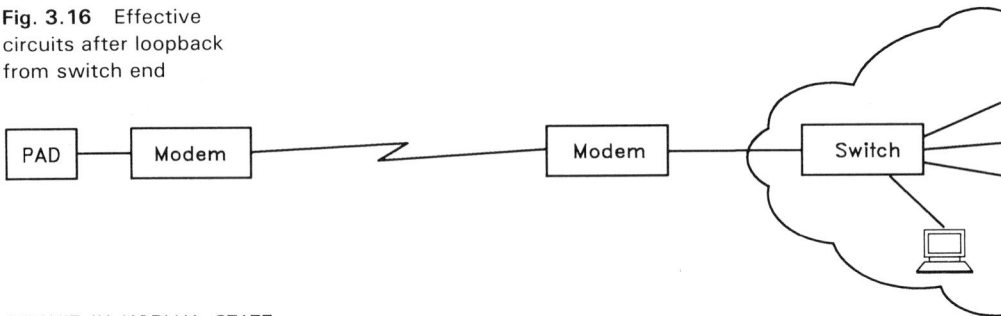

CIRCUIT IN NORMAL STATE

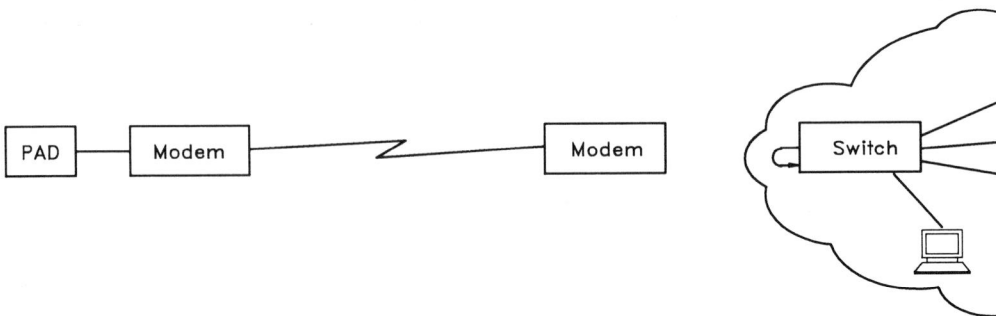

EFFECTIVE CIRCUIT AFTER 'LOOP 1' COMMAND

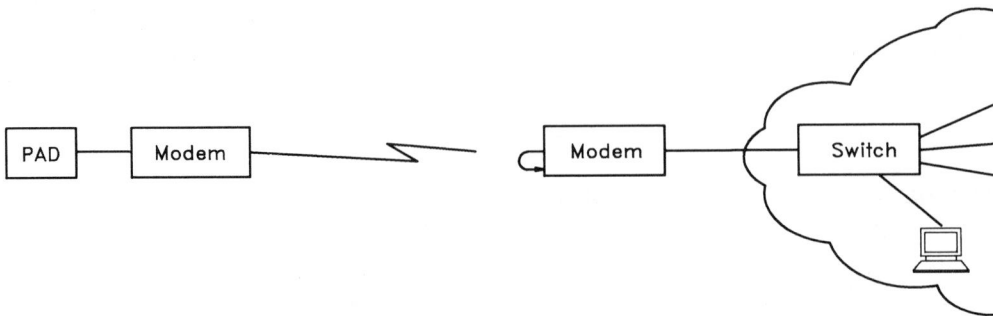

EFFECTIVE CIRCUIT AFTER 'LOOP 3' COMMAND

**Fig. 3.17** Software
loopback commands

```
LOOPBACK

which line?            9
which loopback?        2
wait...
loopback 2 entered
how long to test?      120
test running




RESULTS

which line?            9


test ran for 120 seconds
1500 frames transmitted
1400 frames received correctly
 100 frames received with errors
```

Note that there is a delay after the loopback is requested during which
the modem actually switches into the state and then indicates this to the
controlling software.

## 3.6  Types of network

There are three basic types of network: Public, Private, and Third
Party.

A *public network* is generally one operated by the telephone authority
of the country, and which generally has the same extent as the telephone
system. The authority is referred to as the PTT (Post, Telephone,
Telegraph), and owns the switches and lines, and imposes regulations on
how the network is used. There will also be links to public networks in
other countries, allowing international traffic. The PTT will also operate
their own PADs, and users can rent either terminal lines on the PTT
PAD, or X.25 lines providing a network link for the user's PAD. In
either case the PTT will provide a modem at the user site, linking to the
nearest PTT node. The X.25 links may be used for the user's host or
switch. Fig. 3.18 shows a public network.

The public network only provides connectivity, it does not provide
any computing service. However, the PTT may well provide such value
added services as a separate offering. Value added services would
include electronic mailboxes, links to the Telex system, and various
databases. Such services are all essentially host computers which are
accessed via the public network.

**Fig. 3.18**  Public network

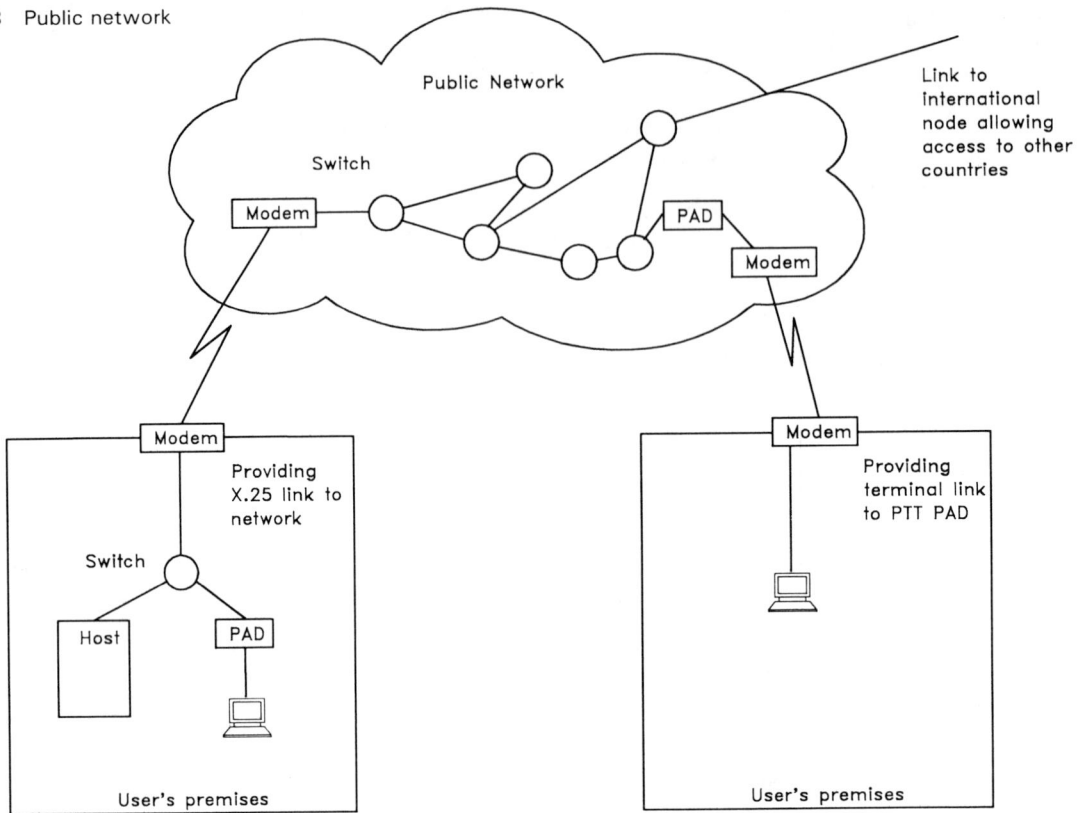

Third party networks take the concept of value added services a step further, and provide their customers with a complete network, some of which may use a public network. Users of third party networks are generally provided with a terminal link, and have access only to the services of the third party. A typical system might be that used by some financial services to provide their customers with up-to-date information. The customers are provided with a terminal, and the third party supplier takes care of all the links back to its host computer. This is shown in Fig. 3.19.

Private networks are much like third party networks but are provided by an organization for its own benefit. The network will often use a hybrid of public network and private links to effect its design, and will usually also be a level more detailed than either the public or third party networks and will address requirements within each user's premises. Fig. 3.20 shows this.

Both third party and private networks require the use of cables to link sites together when the public network is not being used. These links are generally hired from the PTT and are usually priced in such a way that the rental is expensive but there is no charge related to the amount of data carried. This is opposite to the usual pricing of public network

**Fig. 3.19** Third party
network

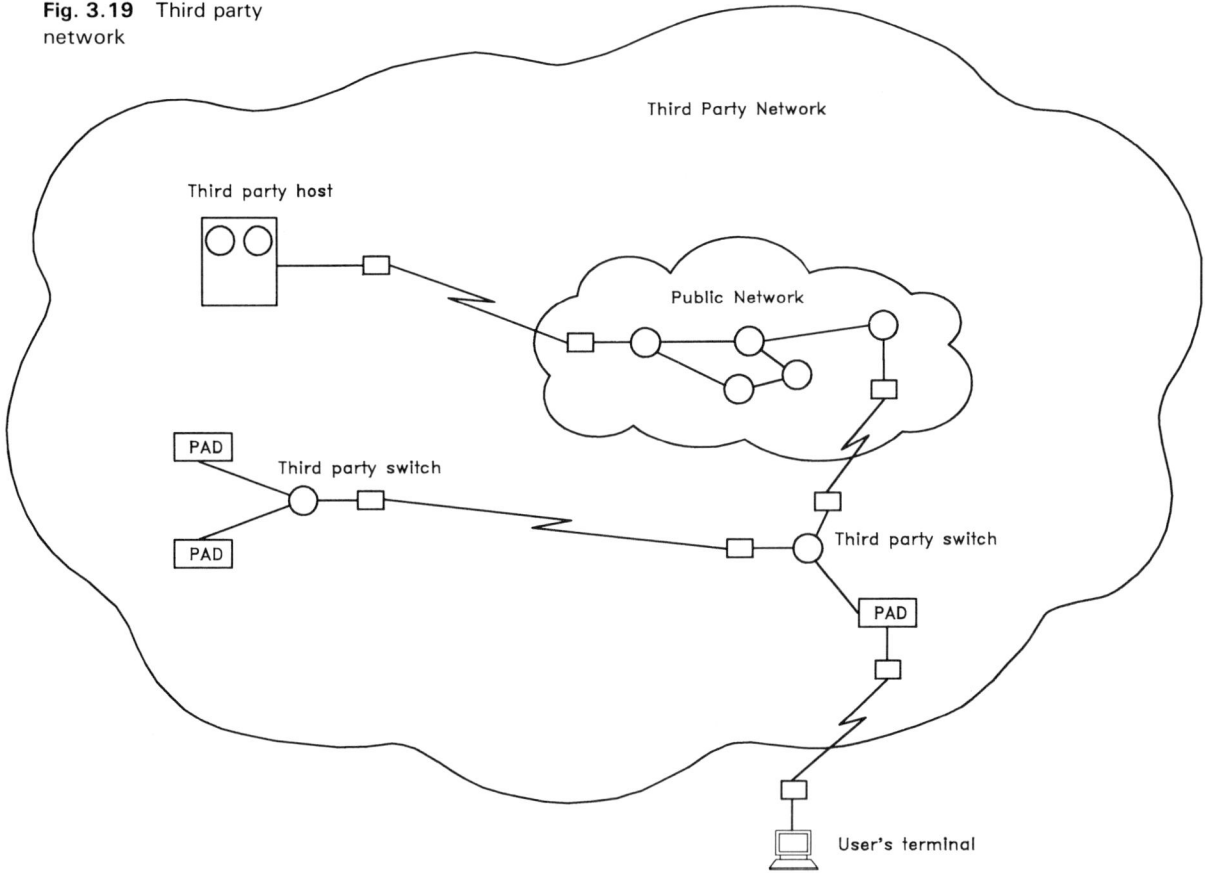

usage, and network designers have to do some careful traffic predictions before deciding which method to use.

For example, a public network charges an installation fee of $A$, an annual rental of $B$, and a cost per packet transmitted of $C$. A dedicated link hired from the same source might have an installation charge of $D$, and an annual rental of $E$. If this part of the network has an expected life of five years then the comparative costs are:

$A + (5 \times B) + (\text{number of packets} \times C)$      for the public network
$D + (5 \times E)$     for the dedicated link

The only variable in this problem – given stable pricing – is the amount of traffic that will be transmitted.

## 3.7 Topologies

Topology is the architecture or shape of a network, and there are a number of classical topologies from which to choose. We have already seen most of these, but a brief list is given here with the salient points of each.

**Fig. 3.20** Private network

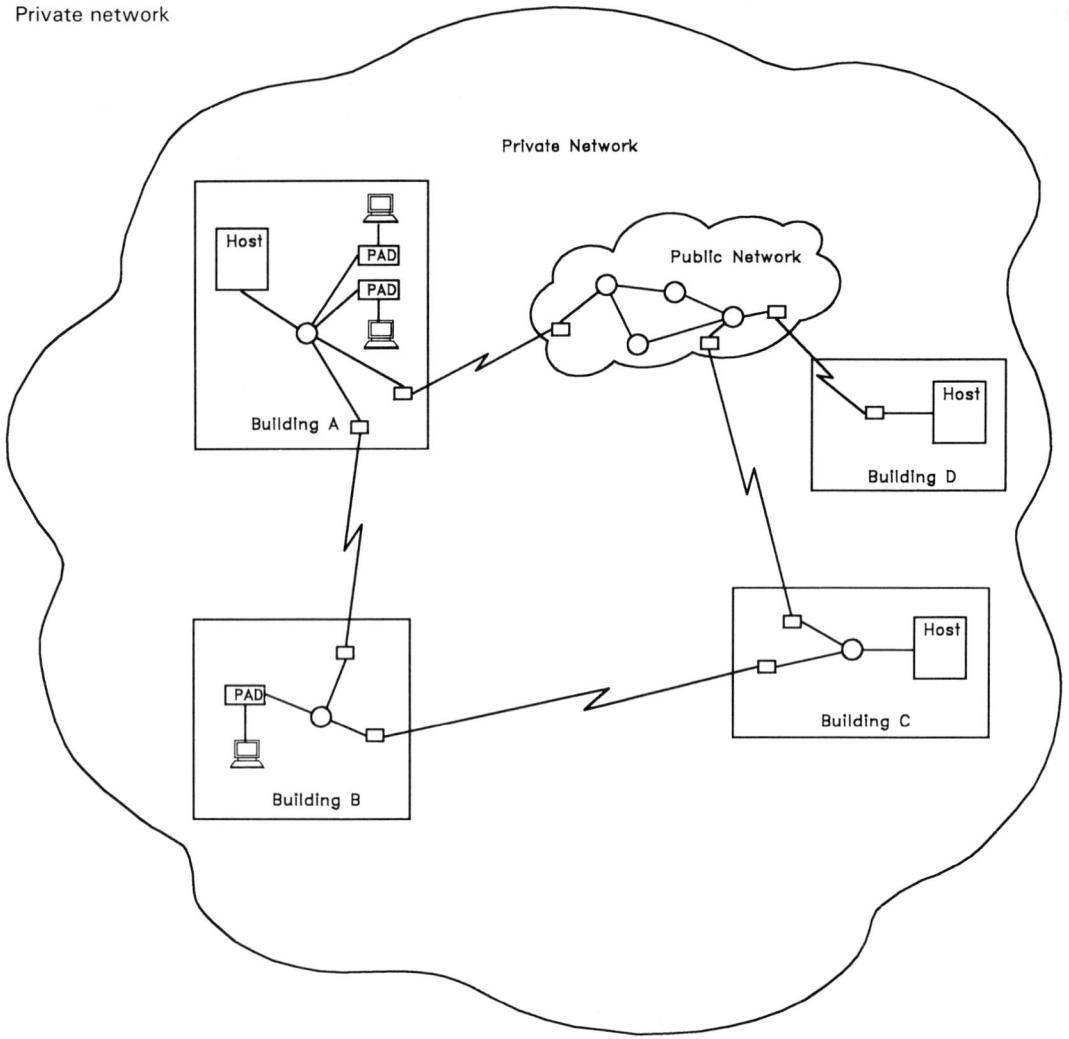

### 3.7.1 Point-to-point and star

This is the simplest and probably the most common way of providing connectivity between two points and is illustrated in Fig. 3.21. When more users are added to a point-to-point network, it grows into the star configuration shown in Fig. 3.22 which was so popular with early mainframes.

This requires no technology above that to provide the basic connection, but suffers from having duplication of cables and interfaces which are probably never all used at the same time.

In the early parts of this book we have seen the disadvantages of this approach. This is especially true when more than one service is involved and when the terminals are microcomputers — as many of them are

**Fig. 3.21** Point-to-point connection

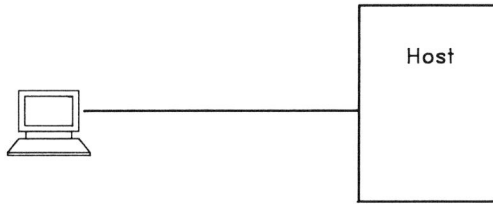

**Fig. 3.22** Star network

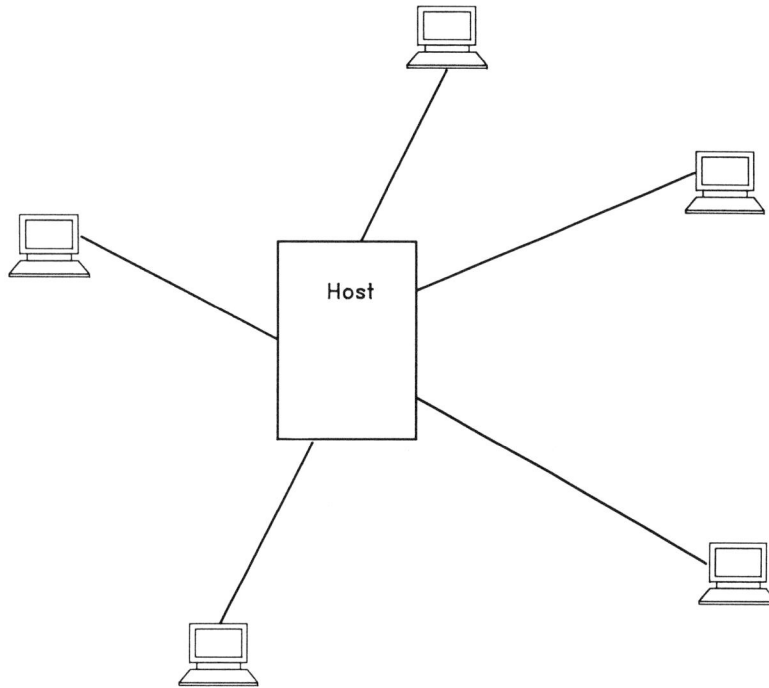

nowadays – which have a legitimate need to exchange data in their own right without having to use the intervening mainframe.

Note that a network may be a star configuration of the sort shown in Fig. 3.23. This looks different to the conventional star network (a host surrounded by users) because of the use of a switch. However, it is a star with all the disadvantages of that topology.

Whilst the central point is merely a switching node and not an end service, it is still a resource that the surrounding units rely on. There will be no problem as long as the switch can cope with all the traffic, but the network designer does need to check the traffic predictions and be sure that the design is suitable. What will happen to terminal traffic if the hosts start to exchange large data files? What happens if the switch goes down? Should the network be augmented with some point-to-point links?

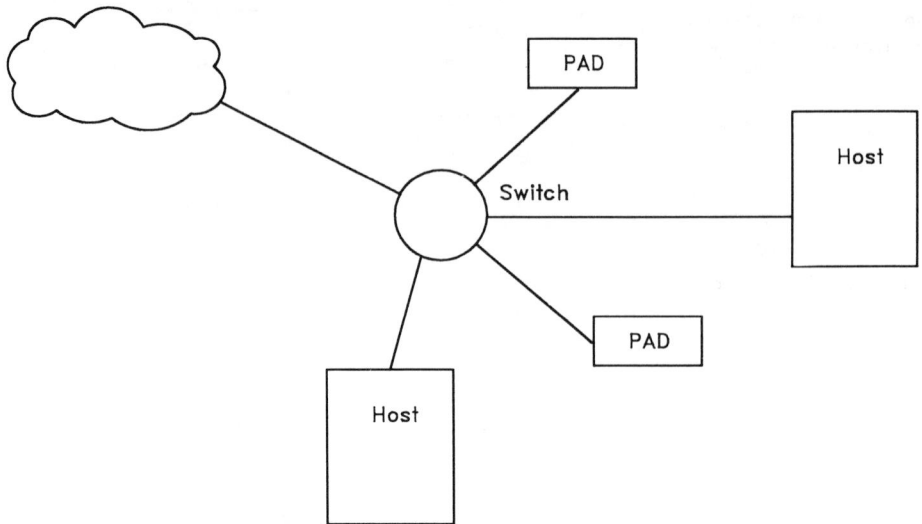

**Fig. 3.23** Star network using central switch

### 3.7.2 Mesh and loop

Mesh is the topology typically used in X.25 networks where nodes are linked on the basis of the amount of traffic expected to flow between them. Thus in Fig. 3.24, B and F are expected to exchange a lot of data, whilst C and E are not and have an indirect linkage.

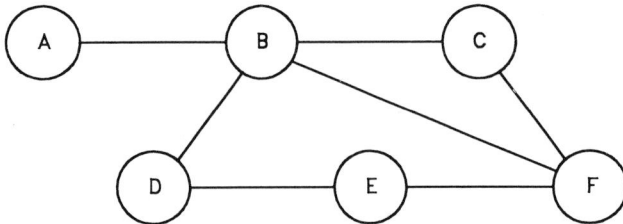

**Fig. 3.24** Mesh network

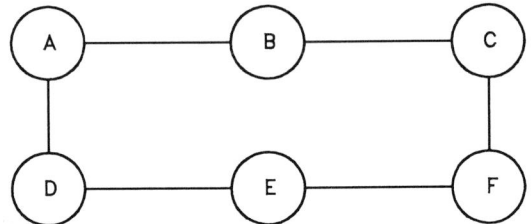

**Fig. 3.25** Loop network

Mesh networks have resilience built-in, and as can be seen in Fig. 3.24 a fault in almost any link still allows data to be transferred between all nodes. This does mean however that the number of cables is higher and therefore may be costly to implement.

By reducing the mesh to a loop as shown in Fig. 3.25 much of the resilience is retained at a probable saving in implementation cost, though of course this relies on the nodes being sufficiently powerful to handle the data. Comparing Figs. 3.24 and 3.25, the loop topology is only an option if node C can handle all the B-F data that would otherwise have its own link.

# 4 Details of X.25

## 4.1 Introduction

Chapters 1 and 2 have shown the basic mechanisms of X.25 including windowing and error recovery. Only a subset of the available frames and packets has been described so far, and this chapter explains the rest of them and where they are used.

The main body of the chapter covers the frames and packets applicable to the 1980 version of X.25. A list of the major differences between this and the 1984 version is included at the end of the chapter. Either version of X.25 may be used as a network protocol, the choice usually being a pragmatic one of what the Network Administration insists upon. Most administrations are committed to a move to the 1984 version.

It may be helpful to read this chapter in conjunction with the frame and packet layouts given in Appendix B.

### 4.1.1 Summary of common frame and packet types

The following frame and packet types were discussed in Chapter 1, and are used for many of the information transfer and flow control procedures of levels two and three.

*Information frame* (Info)
  This is used to carry the level three packets. It carries the send and receive sequence numbers $N(R)$ and $N(S)$ allowing windowing and flow control to be performed.
*Receiver Ready frame* (RR)
  This is normally used to indicate acknowledgement of received Info frames in order to keep the window open. It contains the $N(R)$ value.
*Reject frame* (REJ)
  This indicates an error in the sequence numbering and normally implies that a frame has been corrupted in transit and has been discarded by the hardware. The REJ frame contains the sequence number of the first frame that must be retransmitted, and all frames after this are also retransmitted.

*Call packet* (CALL)
This establishes a Virtual Circuit between the destination and source. These are identified by network addresses. The CALL results in the allocation of LCNs which are used to identify the separate traffic streams in the multiplexed packets.

*Call Accept packet* (CAA)
This is sent by the destination in response to the CALL packet, to indicate that it is willing to participate in the conversation.

*Clear packet* (CLR)
This may be sent at any time to terminate the call and return the resources used by it.

*Clear Confirmation packet* (CLC)
This is sent to acknowledge the CLR.

*Data packet* (Data)
This is used to carry the actual data of the conversation. It carries P(R) and P(S) sequence numbers to provide windowing and flow control.

## 4.2  Other types of frame

The only frames not described in earlier chapters are concerned with starting the link and with closing it down. Before these are described, it is necessary to be a little more precise about the X.25 specification and the functions of the network.

### 4.2.1  Link establishment and end identification

The two ends, or parties, of an X.25 link are identified by the names DTE and DCE. These abbreviations stand for *Data Terminal Equipment* (the user of the network) and *Data Circuit-terminating Equipment* (the access provided by the network). Unfortunately, these names are also used for a different purpose at the electrical level as will be shown in Chapter 7, and it is important to be sure which level is meant when one of the names is used.

X.25 only defines a protocol between a network and a user of it. It does not define the protocol between nodes in the network. This subtlety is shown in Fig. 4.1. Thus to be accurate, an X.25 network is one that implements X.25 for its users. Many X.25 networks are implemented using X.25 protocols between the nodes as we have illustrated in the first two chapters. It is important to realize that this is not a requirement of X.25, and indeed there are disadvantages in doing this.

The network provides an access point – the DCE – for the user – the DTE – to connect to. Once the interface is achieved then the DTE and the DCE can exchange X.25 protocol units. A link between two nodes on a network implemented using X.25 protocol has an arbitrary allocation of DTE and DCE, but there is one of each.

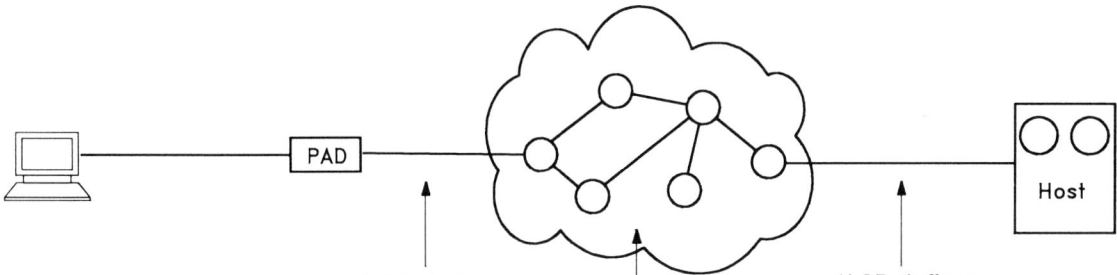

**Fig. 4.1** X.25 defines protocol between the user of the network and the network itself; does not define how data is carried within the network

X.25 defines protocol here

X.25 does not describe what happens in the network

X.25 defines protocol here

Host

PAD

**Fig. 4.2** One end of the link has stopped; what should happen next?

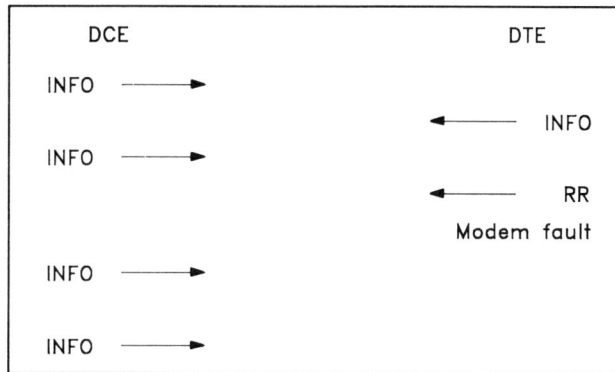

Consider a situation in which the line is operational and frames are being exchanged. One end now stops sending, perhaps because of a modem fault, and the other end continues to send data as shown in Fig. 4.2. So far everything is alright as far as the DCE is concerned, and as shown in Chapter 1 the situation can continue until the window is full. When that happens the DCE can "prompt" the DTE by mechanisms that will be discussed shortly, and will discover that the DTE is not operational when it fails to respond to the prompt. What should the DCE do now? Clearly there is little point in continuing to try to send data since a problem has occurred and the link must be regarded as suspect. What actually happens is that the DCE must try and re-establish contact with the DTE and make the link operational again.

Each end of the link therefore has an operational and non-operational state, and will initialize in the non-operational state. Before the DTE and DCE can exchange frames, it is necessary for them to exchange information to ensure that the link is operational.

The DCE non-operational state is called Disconnected Phase. In this phase it may poll the DTE to solicit a response. One way it can do this is by issuing a DM (*Disconnected Mode*) frame every few seconds. The time between sending DMs is referred to as T1, and is decided by the network administration. This is simply a means of asking: is there anyone there? When the DTE is eventually connected and operational then it will respond to one of the DM frames by sending a SABM (*Set Asynchronous Balanced Mode*). The SABM frame is an indication that the DTE wants to bring the link into operation. The DCE will respond with a UA (*Unnumbered Acknowledgement*) and the link is then operational and Info frames can be exchanged. This is shown in Fig. 4.3.

**Fig. 4.3** Starting the link

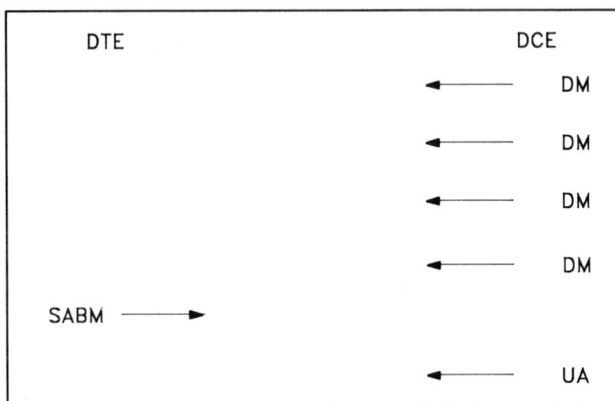

X.25 offers many types of frame and packet for use, but is not always clear on the procedures for using them. The startup situation is one where the recommendation does not say precisely what should happen, and there are therefore some differences and indeed incompatibilities between implementations.

Some implementations poll DISC frames instead of DM frames when in the Disconnected Phase. Such implementations require a UA response to the DISC before a SABM command can be transmitted.

In Disconnected Phase the DCE will respond to frames other than link initialization commands by sending a DM. This is shown in Fig. 4.4.

The DCE may also attempt to start the link by sending a SABM, though this is not usual. The response to a SABM may also be a DM rather than a UA to indicate that the DTE or DCE is unable to operate the link. The component would then remain in Disconnected Phase.

The link may be initiated by a SARM (*Set Asynchronous Response Mode*) rather than a SABM, but use of this mode is declining. This is discussed further in section 4.2.6.

### 4.2.2  Stopping the link

One end may want to stop the link, perhaps to perform an orderly shutdown prior to the component stopping, or in response to a command given by the site manager. The site manager commands were discussed in Chapter 2.

**Fig. 4.4** The DTE tries to send frames before the link has been initialized

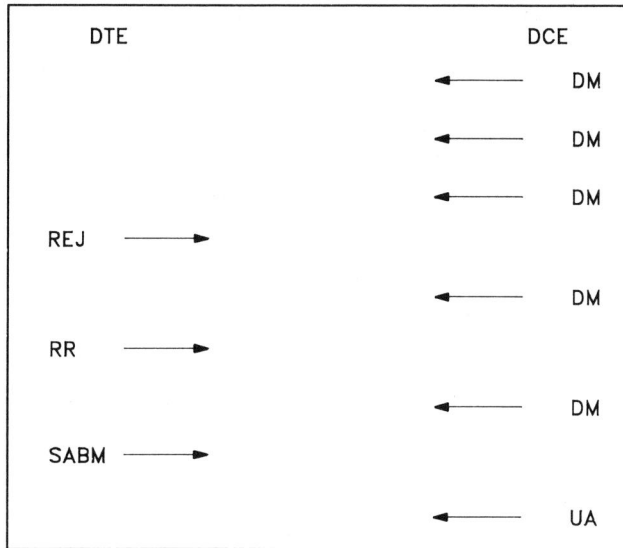

Either end may stop the link by sending a DISC (*Disconnect*) frame. This will be acknowledged with a UA frame. Info frames that have not been acknowledged at the time of the exchange of DISC and UA remain unacknowledged. Having exchanged DISC and UA, the DTE and DCE enter Disconnected Phase. This is usually signified by the polling of DM frames as explained in section 4.2.1.

### 4.2.3 Other frame types

There are two frame types that have not yet been described, the RNR and the CMDR/FRMR.

The RNR (*Receiver Not Ready*) is an acknowledgement sent from one component to another of Info frames that have been received. It is therefore much like the RR frame. However, whereas the RR allows more Info frames to be sent, the RNR indicates a temporary inability to receive more Info frames. This is shown in Fig. 4.5 where the DTE indicates that it does not wish to receive any more frames. Either the DTE or DCE may send an RNR.

**Fig. 4.5** Indication of temporary inability to receive frames

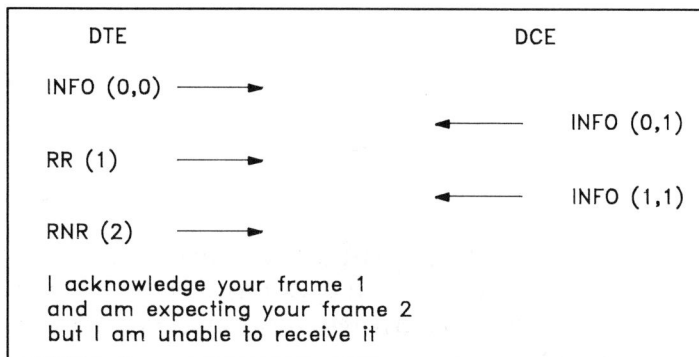

Reasons for sending an RNR include the following:

- Internal buffers are full of previously acknowledged information that has not yet been processed. This may be caused by a terminal controlling the flow of data from a PAD, meaning that the PAD cannot deliver data received from the network.
- Processor too busy to cope. This may be because there is simultaneous activity on many lines of a switch.

When the condition causing the RNR clears, then the component indicates its ability to receive Info frames once more. It may do this in a number of ways, for example:

- The normal method is to send an RR in the same way as if the window is being opened. In the example shown in Fig. 4.5 the DTE would send an RR(2).
- A REJ frame may be sent. This would be most often used if Info frames have been received following the issue of an RNR.
- A SABM may be sent. This will cause a UA response, and both ends of the link will reset the N(R) and N(S) values to zero.

The CMDR/FRMR (*Command Reject/Frame Reject*) frame is used to reject a frame, and indicates that the error cannot be corrected by simple retransmission. This normally results from a protocol error such as that shown in Fig. 4.6. In this situation the DCE is indicating that it is ready for frame 0, whereas this has already been acknowledged. This may be due to some incompatibility between the two X.25 implementations.

**Fig. 4.6** Protocol error resulting in frame rejection

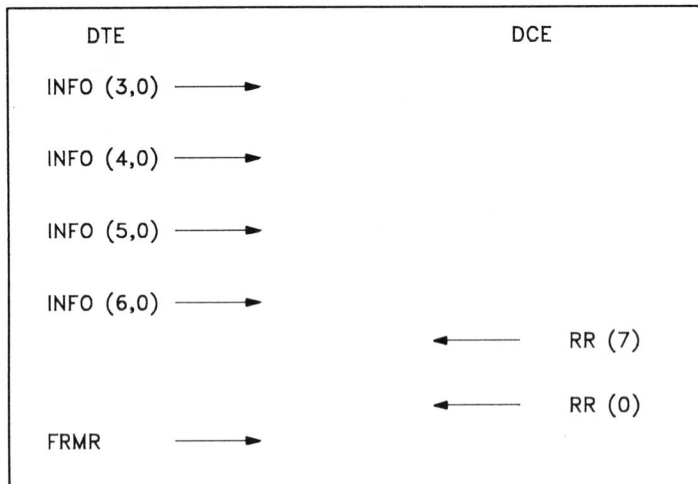

The CMDR/FRMR frame includes indication of the error so that the receiver can deduce the reason for the reject. Part of this information is the send and receive sequence numbers that the sender of the FRMR was expecting.

**Fig. 4.7** Conceptual
FRMR frame layout

| Control field of Rejected frame | Expected N(S) | Expected N(R) | Error type |
|---|---|---|---|

The frame is called a command reject or a frame reject depending on whether the link was started with SARM or SABM commands. Normally the link is started with a SABM and the Frame Reject (FRMR) name is used. Fig. 4.7 shows the conceptual layout of the frame.

### 4.2.4 Commands and responses

Chapter 1 showed that the first byte of a frame is the Address byte. This may contain one of two values referred to as A and B, and indicates whether the frame is a command or a response.

- Frames arriving at the DTE with the address set to A are commands from the DCE.
- Frames arriving at the DTE with the address set to B are responses from the DCE.
- Frames arriving at the DCE with the address set to A are responses from the DTE.
- Frames arriving at the DCE with the address set to B are commands from the DTE.

The full list of commands and responses is shown in Fig. 4.8.

**Fig. 4.8** Frame types

|  | COMMANDS | RESPONSES |
|---|---|---|
| Information transfer | INFO | |
| Supervisory | RR<br>RNR<br>REJ | RR<br>RNR<br>REJ |
| Unnumbered | SARM<br>SABM<br>DISC | DM<br>CMDR/FRMR<br>UA |

Commands are used by one component to force the other component to perform some action. The most common command is the Info frame, which requires some acknowledgement as a response. The response would be one of RR, RNR or REJ, or an Info frame piggybacking the acknowledgement.

The response is not necessarily sent immediately. The receiver of the Info command may wait before acknowledging the information. This action can be modified by the sender setting the Poll bit, which requires

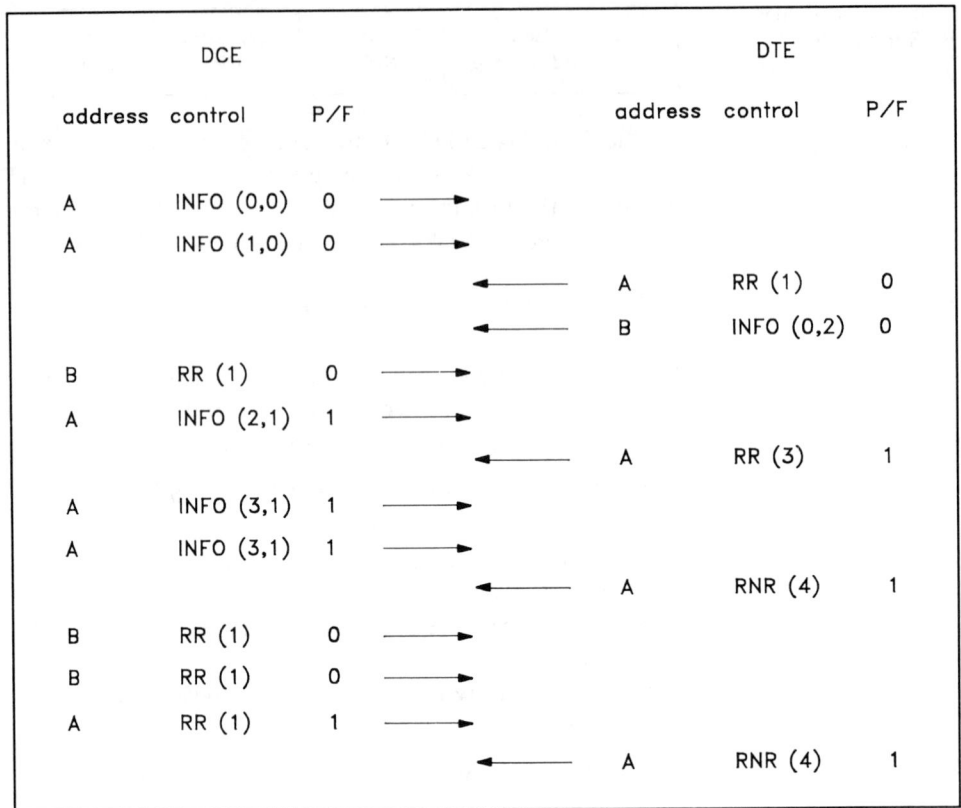

| | DCE | | | | DTE | | |
|---|---|---|---|---|---|---|---|
| | address | control | P/F | | address | control | P/F |
| | A | INFO (0,0) | 0 | ——→ | | | |
| | A | INFO (1,0) | 0 | ——→ | | | |
| | | | | ←—— | A | RR (1) | 0 |
| | | | | ←—— | B | INFO (0,2) | 0 |
| | B | RR (1) | 0 | ——→ | | | |
| | A | INFO (2,1) | 1 | ——→ | | | |
| | | | | ←—— | A | RR (3) | 1 |
| | A | INFO (3,1) | 1 | ——→ | | | |
| | A | INFO (3,1) | 1 | ——→ | | | |
| | | | | ←—— | A | RNR (4) | 1 |
| | B | RR (1) | 0 | ——→ | | | |
| | B | RR (1) | 0 | ——→ | | | |
| | A | RR (1) | 1 | ——→ | | | |
| | | | | ←—— | A | RNR (4) | 1 |

**Fig. 4.9** Use of the Poll bit to demand immediate response to frames

an immediate response from the receiver. The response will have the Final bit set. This is illustrated in Fig. 4.9. Poll and Final bits occupy the same position in the packet; the name depends on whether the packet is a command or a response.

In the first part of this exchange the Poll/Final bits are all set to zero so an immediate response is not required. There is therefore no way of knowing how long this exchange actually took.

The first two Info commands are from the DCE to the DTE, so they carry the address A. The RR from the DTE is a response to the first Info frame so it also carries address A.

The Info frame from the DTE is a command because it requires acknowledgement, and it piggybacks the response to the second Info command from the DCE. Info frames are always considered to be commands rather than responses. Since this frame is sent from the DTE the address is B. This is acknowledged by the RR response from the DCE with address B.

The DCE then sends an Info command with the Poll bit set to demand an immediate response. In this case the response is an RR and the Final bit is set to indicate that this is a response to a polled command. Both addresses are A because the frames are a command from, and a response to, the DCE.

The DCE then sends another polled Info command but this does not receive an immediate response. In fact the response must be received within time T1 of sending the poll, and this time is defined by the network administration. The DTE may not be responding because the command frame was lost in transmission. This would normally be detected by the sequence number mechanism in combination with a REJ frame. However, because this is a polled command, the DCE needs to sort out the problem without delay.

The DCE retransmits the polled frame to try and gain a response. In this case a response with the Final bit set is received. The polled command will actually be repeated up to N2 times before the sender assumes that a fault exists. The sender will then try to recover the link by sending a SABM or DISC. The number N2 is defined by the network administration.

Two RR responses are then sent by the DCE. They are responses to the original Info command sent by the DTE. There is no limit on sending Supervisory frames to indicate that the sender is in a particular state. The RR command that follows these is a request for the state of the other end, and is used to reassure the sender that the link is still operational. It is not necessary to retransmit a polled Info frame. A polled RR could be sent instead to try and gain a response, and this may be cheaper and more efficient since the frame is shorter.

Figure 4.10 shows a further example. The DCE sends a polled Info command and gets an RR response with the Final bit set. The DTE then

**Fig. 4.10** Further example of use of Poll bit, showing necessity for the address field

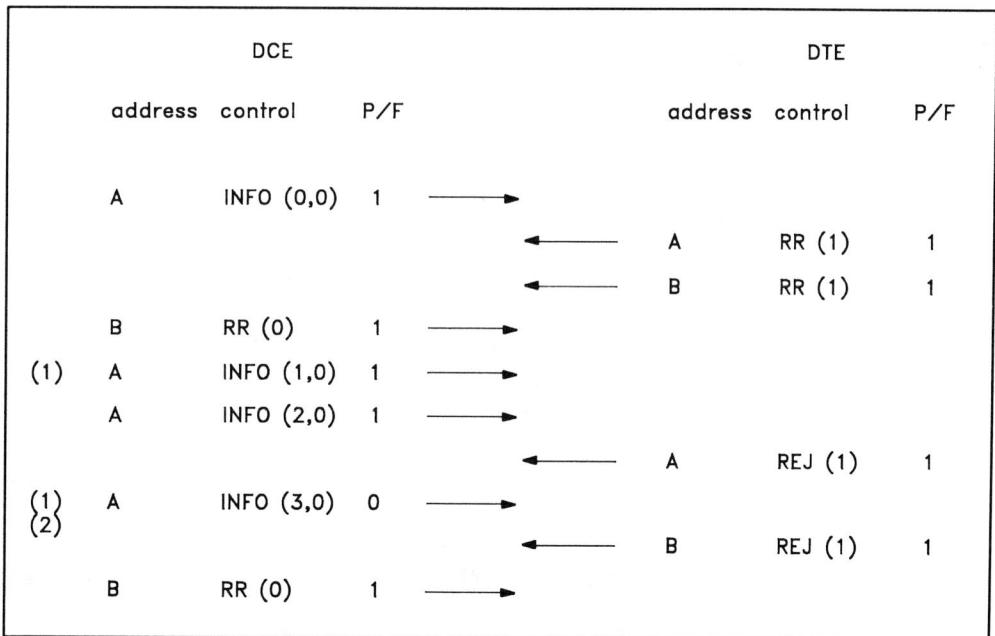

| | DCE | | | | DTE | | |
|---|---|---|---|---|---|---|---|
| | address | control | P/F | | address | control | P/F |
| | A | INFO (0,0) | 1 | → | | | |
| | | | | ← | A | RR (1) | 1 |
| | | | | ← | B | RR (1) | 1 |
| | B | RR (0) | 1 | → | | | |
| (1) | A | INFO (1,0) | 1 | → | | | |
| | A | INFO (2,0) | 1 | → | | | |
| | | | | ← | A | REJ (1) | 1 |
| (1)(2) | A | INFO (3,0) | 0 | → | | | |
| | | | | ← | B | REJ (1) | 1 |
| | B | RR (0) | 1 | → | | | |

Notes: (1) This frame lost.
(2) Timeout occurs and retransmitted frames 1, 2, and 3 lost.

requests the status of the DCE by sending a polled RR command, and receives an RR response. Info frame number one from the DCE is lost in transmission so frame number two gets a REJ response. Because frame two was polled, the REJ has the final bit set.

Frame three is then sent by the DCE. This is theoretically an error, but it may be caused by the DCE sending a correct frame – as far as it was concerned – whilst the DTE was transmitting the REJ. It must be remembered that there is a data path in each direction and the two ends can transmit simultaneously. Having sent a REJ the DTE will ignore all Info frames until it receives the one it is waiting for.

The DCE responds to the REJ by retransmitting frames one, two, and three; however, these are lost in transmission. The DCE – as far as it is concerned – has done everything correctly. The DTE however is still trying to receive frame one. The DTE sends a polled REJ asking for the state of the DCE and receives an RR in response. The DCE will then retransmit the requested frames.

Notice in Fig. 4.10 that the address field is a necessary part of the frame to distinguish commands and responses from the RR, RNR, and REJ. Information and unnumbered frames are always one or the other so the address field is usually redundant. Actually DM is identical to a SARM except for the address field as shown in Appendix B.

The address field is eight bits wide and is encoded 3 (00000011) for A and 1 (00000001) for B. Frames with other values will be ignored. Appendix B shows detailed frame layouts.

### 4.2.5 Timers and numbers

The timers and numbers used with the link layer of X.25 are as follows:

*Timer T1, Retransmission timer* This is the time after which a frame is retransmitted. It is most often used with polled frames, but it may also be used with other frames. The time should be greater than the sum of the following:

- The time to send a frame. The longest frame is an Info frame and the maximum length is derived from the value of N1. (See below.)
- The time for the frame to propagate through the modems and wires to the other end.
- The time the other end waits before acknowledging the frame. This is defined by T2.
- The time to transmit the acknowledging frame.
- The time for the acknowledging frame to propagate across the link.

*Timer T2, Acknowledgement timer* This is the maximum time that a component can wait before sending an acknowledgement to a frame that it has received. It is used to respond before timer T1 expires at the other end, but still allows time for data to arrive at this end to be transmitted as Info frames, so that RRs are not sent unnecessarily. T2 must be less than T1.

*Number N1, Maximum number of bits in an Info frame*   This is the largest possible size of Info frame that it is possible to send over the link.

*Number N2, Maximum number of transmissions*   This is the maximum number of times a frame will be transmitted. Following the initial transmission and the expiry of T1, a further (N2 − 1) retransmissions and expires will occur.

*Number K, Window size*   The maximum number of unacknowledged Info frames that may be outstanding. This will be less than the modulus which is normally eight.

Typical values are as follows:

| T1 | 5 sec | N1 | 8000 bits | K | 7 |
|----|-------|----|-----------|---|---|
| T2 | 1 sec | N2 | 20 | | |

### 4.2.6  Link modes

The link may run in one of two modes: LAP (*Link Access Protocol*) or LAPB (*Balanced*), depending on whether the DTE starts the link with a SARM or a SABM frame. LAPB is used almost exclusively and is the procedure described in the rest of this chapter.

LAP operation has the following differences:

- Having received a SARM and responded with a UA, the DCE will transmit a SARM and wait for a UA. The two directions are thus initialized independently.
- Either end may send a SARM to re-initialize its sending direction, and only frames in that direction start again with an N(S) of zero.
- The link is closed by each end transmitting a DISC to stop the two directions.
- During Data Transfer phase, the DCE will not transmit RR, RNR, or REJ command frames.

## 4.3  Other types of packet

All the frame types have now been examined, and the function of this section is to examine all the types of packet. As explained earlier, this section looks at the packets defined in the 1980 version of X.25. The 1984 version is described at the end of the chapter. Normally the network administration will dictate which version is used, and this will normally be the same as the layer two version.

## 4.3.1 Logical channel assignment

In Chapter 1 it was stated that the DTE and DCE agree on a range of LCNs that may be used for calls going over the link. An annex to the X.25 recommendation suggests a structure for the numbering as shown in Fig. 4.11:

● LCN 0 is not valid. It should be noted that many administrations ignore this and use LCN 0.
● "Incoming only" channels may only be used by calls initiated from the network side. The DTE may not send a call packet using any of these LCNs.
● "Bothway channels" may be used for any calls.
● "Outgoing only" channels may only be used by calls initiated from the DTE. Calls from the network side may not be established using any of these LCNs.

**Fig. 4.11** Structure of LCN allocations recommended in X.25

Thus, four ranges of channel numbers are agreed between the DTE and

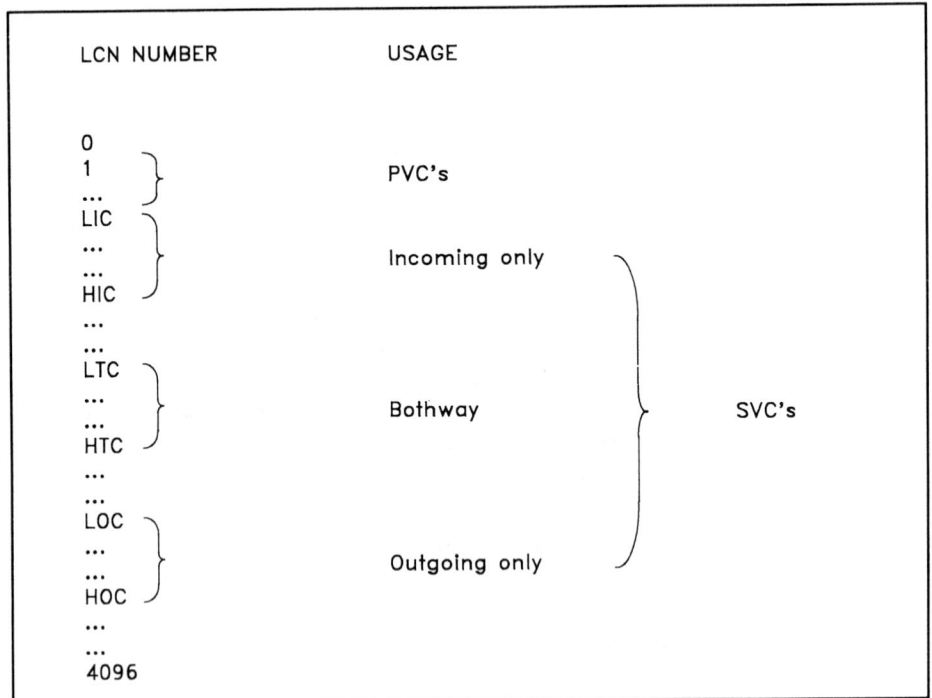

```
LCN NUMBER              USAGE

0
1              }        PVC's
...
LIC        ⎫
...        ⎬            Incoming only
...        ⎭
HIC
...
...
LTC        ⎫
...        ⎬            Bothway                    SVC's
...        ⎭
HTC
...
...
LOC        ⎫
...        ⎬            Outgoing only
...        ⎭
HOC
...
...
4096
```

LIC — Lowest Incoming Channel
HIC — Highest Incoming Channel
LTC — Lowest Bothway Channel
HTC — Highest Bothway Channel
LOC — Lowest Outgoing Channel
HOC — Highest Outgoing Channel

DCE, one for PVCs and three for SVCs. The three SVC ranges may be needed as follows:

- A supplier of a value added service may have all LCNs on a public network used by incoming calls to the service being provided. A number of outgoing only LCNs would be needed to ensure that the supplier could call out to the network.
- A supplier of networking equipment may have all LCNs on a public network used by outgoing calls to test out PADs. A number of incoming LCNs would be needed to ensure that customers could call into the suppliers' database.

Using the various LCN ranges a suitable configuration can be chosen for the application.

In most cases the user of the network service will indicate the desired LCNs when the application to use the network is made. The network administration will then configure the DCE and inform the user of the LCN numbers to be used. The user must then configure the DTE if this has not been supplied by the administration. In private or local area networks where the user has both DCE and DTE to configure then the allocation of LCNs is entirely the choice of the user.

It must be remembered that an incoming call at one end of the link is an outgoing call at the other end. The ranges would therefore be configured as shown in Fig. 4.12.

**Fig. 4.12** Outgoing calls at one end are incoming calls at the other

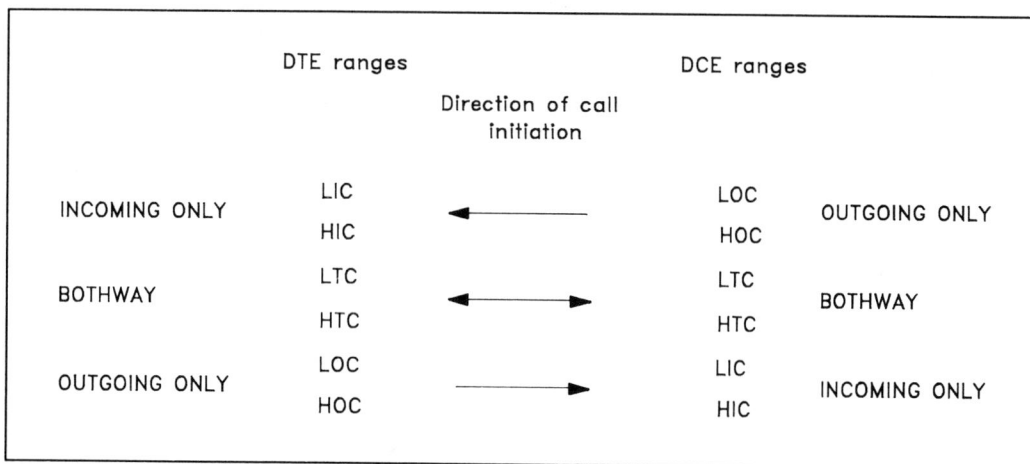

It is possible but not common to use the LCG (*Logical Channel Group*) of the LCI to indicate boundaries of the ranges. This could be arranged as follows as is done by the British PTT.

| *Groups* | *Possible LCNs* | *Use* |
|---|---|---|
| 6 and 7 | 600-7FF | "Outgoing only" (from DTE) |
| 4 and 5 | 400-5FF | "Bothway" |
| 2 and 3 | 200-3FF | "Incoming only" (to DTE) |
| 0 and 1 | 001-1FF | PVCs |

Thus a user anywhere on the network subscribing to five bothway LCNs would be allocated LCNs 400 to 404.

If both the DTE and the DCE initiate a call simultaneously by transmitting a call packet, then if they both use the same LCN there will be a collision. In this case the DCE resolves the conflict by issuing a Clear Request back to the remote caller. However, it is better if the conflict, and therefore the delay, can be avoided.

The X.25 recommendation suggests that the DCE uses LCNs from the bottom of the incoming-only (with respect to the DTE) range first, moving up to the top of the range, and then using bothway LCNs from the bottom to the top of the range. The DTE should allocate from the top of its outgoing-only range first, moving down to the bottom of the range, and then using bothway LCNs from the top to the bottom of the range.

### 4.3.2 Initializing the packet layer

Once layer two has been initialized by the exchange of SABM and UA, then layer three will go through a similar process so that the DTE and DCE synchronize their packet activities. At this startup time, the only real effect of the exchange is to have a confirmatory handshake and go through internal procedures to initialize LCN tables. If the packet level initialization follows an error then the effect is serious since all calls in progress will be lost.

The layer three initialization is normally performed by the DCE in this startup situation, and is achieved by sending a *Restart Indication* packet (RES). The DTE will respond with a *Restart Confirmation* (REC). This is shown in Fig. 4.13.

Calls can now be made. If the DTE fails to respond with a REC

**Fig. 4.13** Starting the packet layer activities

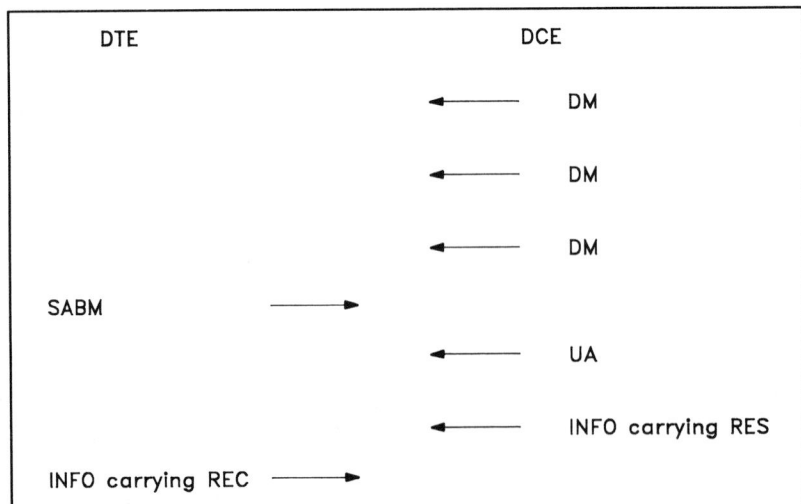

100

within a set time, then normally the DCE will repeat the RES until a response is made. The time is set by the network administrator and is referred to as timer T10. This is normally around sixty seconds.

Restarts may be performed at any time and by either the DTE or DCE. There are two main reasons for using restarts:

- *Local Procedure Error*  An incorrect packet has been sent or some other violation of the layer three protocol has occurred.
- *Network Congestion*  Either the DTE or DCE has too many calls in progress and wishes to terminate them all, or, by implication, if they are not terminated then corruption of the data is possible.

When the DTE and DCE exchange RES and REC then all calls will be cleared by Clear Request packets being issued by the DCE to the remote ends. Should both the DTE and DCE transmit RES at the same time then the exchange is complete and no confirmations are issued.

The RES packet always contains a Restarting Cause field which indicates which of the two reasons shown above has brought about the restart. There is a third reason which is used in the startup situation to indicate that the network is now operational.

When issued by the DTE, the RES may optionally contain a Diagnostic Code, which gives further details of the problem and which is passed on to the remote DTE in the Clear Request. The Diagnostic Code is mandatory when the DCE issues the RES; and its possible values are given in the X.25 recommendation. These are summarized in Appendix B.

Since the RES and REC are layer three packets, they contain fields for LCG and LCN as shown for other packet types in Chapter 1. These packets refer to all calls rather than specific ones, and thus the LCG and LCN are set to zero.

### 4.3.3  Unrecoverable errors

The Restart mechanism implements a complete re-initialization. There is an optional mechanism which allows the DCE to send to the DTE an indication of an error condition, so that the DTE can attempt to resolve the problem without clearing all the calls. This is called the *Diagnostic packet*. It is typically issued when the DCE receives packets that are not structured properly, or which have an unassigned LCN.

### 4.3.4  Within the call

The procedures for setting-up and clearing the call have been described in Chapter 1. An example call is shown in Fig. 4.14 for reference. In this example the DTE has initiated the call, but the Call Request could also have been transmitted by the DCE, having received it from a remote DTE.

```
           DTE                                    DCE

LCN   C    P(S)  P(R)   Data        LCN   C    P(S)  P(R)   Data


400   CALL              ──────────►

                              ◄──────────  400   CAA
                              ◄──────────  400   DATA  0     0
                              ◄──────────  400   DATA  1     0

400   RR           2    ──────────►
400   DATA  0      2    ──────────►

                              ◄──────────  400   RR          1
                              ◄──────────  400   CLR

400   CLC               ──────────►
```

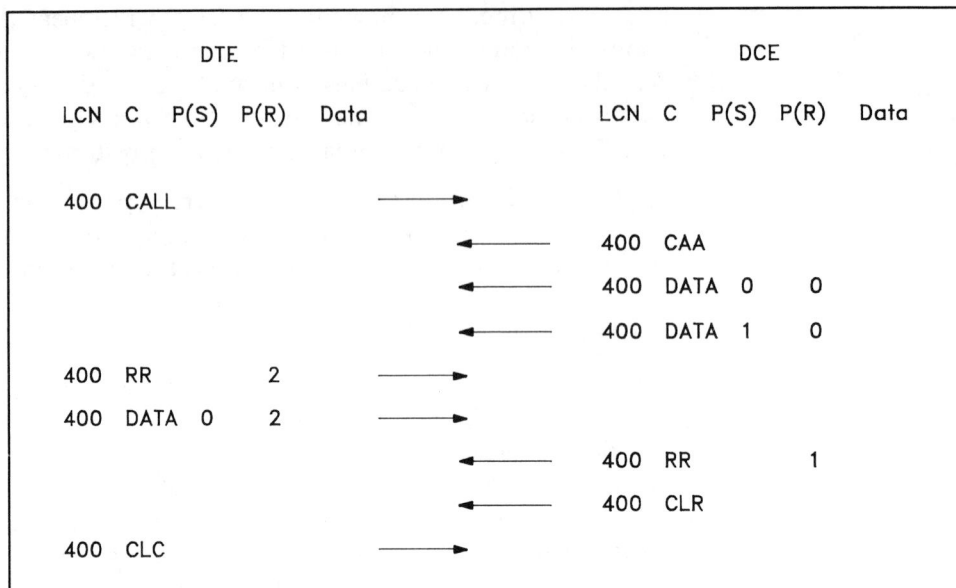

Fig. 4.14 An example of a call showing the layer three packets

**Flow control**  The RR packet is used as an acknowledgement of packets that have been received, and keeps the window open so that more packets may be sent. The standard window size is two, though this may be modified in the Call packet, as shown below, for any call.

The packet numbering is always modulo-8 unless extended operation is available. This option implements numbering with modulo-128, with a corresponding increase in possible window sizes. Extended operation is typically used on parts of the network involving satellites or other slow links. In such cases, where a great number of packets may be awaiting acknowledgement, extended operation improves throughput reduction due to delays in the transmission.

RNR packets may be transmitted by the DTE or DCE to indicate a temporary situation where they are unable to receive any more packets. This is normally due to the transmitter of the RNR having received a flood of packets which have to be processed before any more can be accepted. The transmitter will later send an RR to indicate that more data can be sent. This is shown in Fig. 4.15, where the DTE is temporarily busy.

**The Call packet**  The Call packet contains the following fields:

- *LCG and LCN* (i.e. the LCN to be used for this call between this DCE and DTE)  The LCN only has local significance (it is used between the local DTE and DCE) and therefore the remote DTE and DCE will, in general, use a different LCN. This is also true of intermediate components of the network if this is implemented using X.25 protocol.

102

**Fig. 4.15** Use of packet level RNR to indicate a temporary inability to receive packets

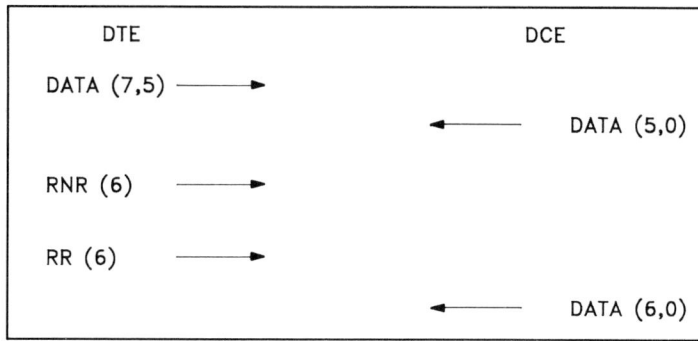

```
         DTE                              DCE

    DATA (7,5) ──────▶
                              ◀────────── DATA (5,0)


    RNR (6)    ──────▶


    RR (6)     ──────▶

                              ◀────────── DATA (6,0)
```

- *Called address and length of the address*   This is an identification of the remote DTE to which the call is being made. X.25 addresses are defined in X.121 to be a maximum of 14 digits long (see Chapter 6) though the length field in the Call packet is four bits long, allowing fifteen digit addresses to be used. On public networks the address format has a fixed structure which is explained in Chapter 6. Private networks may use any structure including the use of zero length addresses if this can be handled by the equipment.
- *Calling address and length of the address*   This is an identification of the local DTE making the call. It informs the remote DTE, and the components in the network, of the calling party. The local DCE may set this address automatically, particularly if it is a public network, but some networks insist that it is set correctly by the DTE.
- *Facilities and length of facilities field*   This is a set of requests made by the DTE either of the network or of the remote DTE. It is discussed below.
- *Call User Data*   This optional field is simply data that is passed between the DTE at the local end, and the remote DTE. Its usage may be defined by a higher-level protocol such as X.29. The field may be up to 16 bytes long.

The scheme recommended in X.25 and X.29 for Call User Data is to split the field into two sub-fields.

The first sub-field is four bytes long and is designated the protocol identifier. The first two bits show the protocol type as follows:

| | |
|---|---|
| 00 | CCITT defined |
| 01 | National use |
| 10 | International use |
| 11 | DTE-to-DTE use |

The only standard encoding of the first byte is 0000 0001 indicating X.29. The use of the remaining three bytes is reserved in the recommendation and they must be set to zero.

The second sub-field is up to twelve bytes long and contains text. This is normally text sent from the PAD to the host indicating additional routing information not carried in the address. The meaning is defined by the host. For example, the called address may specify the remote DTE, and a particular machine on the local network at the remote site, whereas the User Data might specify a particular class of communications software. The sub-field is usually set by the user of the PAD when a call request is made.

For example, a PAD command of

23425871278604, D = FAST

would result in a Call User Data field of

01  00  00  00  46  41  53  54      (hex)
(X.29)              F   A   S   T      (ASCII)

and might indicate that the user required a faster response from the software, and by implication was prepared to pay more for it.

The scheme recommended for X.29 is sometimes extended to cater for other protocols not recognized by the standards body. An example of this is the "Green Book" protocols mentioned in Chapter 2.

The facilities length field is six bits long, allowing up to 63 bytes of facilities. If the length is zero then the facilities field is absent and the length is immediately followed by Call User Data if present, or by the end of the frame. The available X.25 facilities are shown in the next section.

**Facilities**   A number of facilities are made available to the DTE by the network administration. These facilities have to be agreed between the network administration and the network user, and once agreed are always available to the DTE. The facilities often incur additional charges over and above the normal rental. Since facilities are agreed when the user subscribes to the network, they are often referred to as subscription options. The facilities are as follows:

*Extended packet sequence numbering*
    This changes the packet numbering from modulo-8 to modulo-128. This therefore extends the field required for sequence numbers and the structure of the packets is different.
*Non-standard default window sizes*
    Normally the default window size is two, and a call may select a window size of between one and seven. If this facility is available, then the user chooses a different default at subscription time which applies to all calls. Again, individual calls may choose a different value than the default. Some network administrations may permit different window sizes in each direction.

*Default throughput class assignment*

The network may provide throughput classes, which are defined rates of data transfer that the call can achieve. Without this facility then a default is assigned by the network administration, which may be altered for individual calls. If this facility is available then the user chooses a different default at subscription time which applies to all calls. Again individual calls may choose a different value than the default. Some network administrations may permit different throughput classes in each direction.

*Packet retransmission*

Chapter 1 showed the procedure for layer two Reject frames, which allows either end to request retransmission of frames following a sequence error. Normally, such a feature is not available at layer three, and use of the Reject packet is not allowed. This feature is not a necessity since layer two should deliver packets in sequence, and any more serious problems can be sorted out by resetting the call. (Resetting is described later in this chapter.) If packet retransmission is selected at subscription time then the DTE may use the Reject packet to request retransmission of packets not already acknowledged by an RR packet. This facility might be useful in some DTEs where memory is very limited such that there is no room for packets delivered by layer two, and the DTE cannot cope with a full window.

*Incoming calls barred/Outgoing calls barred*

These facilities control the ability to initiate calls in the specified direction relative to the DTE. Such controls apply to all LCNs. Two equivalent facilities are available, called One-way Logical Channel Outgoing and One-way Logical Channel Incoming, which apply to a range of LCNs. This is a rather more positive way of achieving the effect than by using LCN ranges as shown earlier in this chapter.

*Closed User Group*

This facility may be subscribed to by a number of DTEs that may then communicate with each other. Normally no calls into or out of the group are permitted, and this may be useful in applications where sensitive data is present. Further facilities are available to allow calls into or out of the Closed User Group. A special case is the Bilateral Closed User Group where only two DTEs belong to the group and normally communicate only with each other.

*Reverse Charging*

If this facility is subscribed to then the DTE is allowed to request Reverse Charging in Call Requests transmitted to the DCE. If such calls are accepted by the remote DTE, then the remote DTE is charged for the network resources used during the call.

*Reverse Charging acceptance*

If this facility is subscribed to then the DCE will allow incoming Call Request packets to the DTE which specify Reverse Charging. The DTE may reject such calls, by issuing a Clear, or accept them. If the DTE accepts the call then the user is subsequently charged by the network administration for the resources used during the call. If this facility is not subscribed to then the DCE will not present Reverse Charging requests to the DTE, and the DCE will issue a Clear back through the network to the calling DTE.

*RPOA selection*

If this facility is subscribed to then a Call Request packet can request use of *Recognized Private Operating Agencies* (RPOA). This allows the DTE to send calls into the network which are then routed to a private network. This is achieved by the use of a *Data Network Identification Code* (DNIC) specifying the agency. The DNIC is discussed in more detail in Chapter 6.

*Non-standard default packet size*

Normally the default packet size is 128 bytes, though a call may select a different packet size. The packet size specifies the size of the data field in a data packet, and does not include protocol fields such as the P(R) and P(S) values. If this facility is subscribed to then the user chooses a different default at subscription time which applies to all calls. Again, individual calls may choose a different value than the default. Some network administrations may permit different packet sizes in each direction.

*Flow control parameter negotiation*

Flow control parameters are the window size and packet size. If this facility is subscribed to then individual calls may request a different value for their flow control parameters than the current default. If the facility is not subscribed to then all calls will use the default values whether these are standard or non-standard.

Outgoing calls from the DTE that do not request specific values will always use the current defaults for flow control parameters whether negotiation is available or not. If negotiation is available then the DTE may request different values for each direction if this is allowed by the network administration.

If negotiation is available then all Call Requests transmitted by the DCE will indicate values for the window and packet sizes in each direction. The DTE can negotiate different values than those by indicating the desired new values in the Call Accept packet. The new negotiated values are then used for the call across the link. The values negotiated by the DTE must conform to the following rules:

- If the requested window size is one, then the negotiated value must be one or two.
- If the requested window size is greater than one, then the negotiated value must be between the requested value and two, inclusive.
- If the requested packet size is less than 128, then the negotiated value must be between 128 and the requested value, inclusive.
- If the requested packet size is 128 or above, then the negotiated value must be between 128 and the requested value, inclusive.

If negotiation is available and the DTE transmits a Call Request, then the Call Accept returned by the DCE may negotiate different values for the flow control parameters. The rules for negotiations are the same as for the DTE.

There is no relationship between the values of flow control parameters at the two DTE/DCE links. If they are different then the network must resolve the differences, perhaps by segmenting or combining packets. Likewise, if the network is implemented using X.25 protocol then different values may be used for each link utilized by a call. Allowed packet sizes are 16, 32, 64, 128, 256, 512, or 1024 bytes. An implementation may use any subset as long as it includes 128.

*Throughput class negotiation*

If this facility is subscribed to then individual Call Requests may request values of throughput class. The value requested by the original calling DTE may be negotiated in a downward direction by its DCE, or by the called DTE, in the Call Accept.

*Fast Select*

If this facility is subscribed to, then it enables the DTE to use the Fast Select facility in Call Requests. This has a number of effects:

- The maximum length of the Call User Data field in the Call Request packet is increased from 16 to 128 bytes. This allows the calling DTE to pass considerably more information to the called DTE about the nature of the call.
- The Call Accept packet is extended to contain 128 bytes of Call User Data, allowing information to be transferred back to the calling DTE.
- The Clear Request packet is extended to contain two address fields and lengths as in the Call Request, to indicate the network addresses of the two DTEs. The function of these addresses is not defined and the fields must not be used – the lengths are both set to zero.
- The Clear Request packet is extended to contain a facilities and a facility length field as in the Call Request. The function of the facilities is not defined and the field must not be used – the length is set to zero.
- The Clear Request packet is extended to contain a Clear User Data field of up to 128 bytes. This is used by the two DTEs to exchange data when the call is cleared.

All Call User Data fields must adhere to the recommendation already described for the Call Request, that is the first four bytes must be used to indicate the protocol of the call. The Clear User Data has no recommendations associated with its use.

When used as indicated above the Fast Select facility essentially just allows extra data to be communicated between the DTEs during the call setup and Call Clearing phases. This is known as normal or non-restricted response. It is most often used with higher-level protocols to pass extra connection information between the two parties.

An alternative is Fast Select with restricted response, and in this procedure the only response allowed to the Call Request is a Clear Request. Thus the entire exchange consists of the transmission of a Call Request and a Clear Request, each containing up to 128 bytes of User Data. This procedure is often known as a Minicall and is typically used by network management systems to poll data. The network management host transmits a Call Request with User Data encoded to say: "Give me data on ports 7, 8 and 19". The unit receiving the call — perhaps a PAD — would return a Clear Request with the status of the ports encoded in the User Data. Such a use of the minicall is likely to be expensive in terms of network usage if the polling is frequent.

The ability to receive fast select calls from the network is a separate facility to the ability to initiate such calls.

**Summary of facilities**  The facilities available from the network administration described above are summarized below.

*Extended packet sequence numbering*
Permanently changes packet numbering from modulo-8 to modulo-128

*Non-standard default window size*
Permanently changes default window size.

*Default throughput class assignment*
Permanently changes default throughput class.

*Packet retransmission*
Permanently allows use of Reject packet.

*Incoming/outgoing calls barred*
Permanently disallows Call Requests.

*Closed User Group*
Permanently assigns membership of Closed User Group.

*Reverse Charging*
Permanently allows the DTE to request reverse charging on individual calls

*Reverse Charging acceptance*
Permanently allows the DCE to transmit reverse charge Call Requests to the DTE.

*RPOA selection*
Permanently allows use of Recognized Private Operating Agencies.

*Non-standard default packet size*
Permanently changes default packet size.

*Flow control parameters negotiation*
Permanently allows calls to request and negotiate window and packet sizes.

*Throughput class negotiation*
   Permanently allows calls to request and negotiate throughput class.
*Fast Select*
   Permanently allows calls to request Fast Select.

The provision of these facilities depends on the capabilities of the equipment operated by the network administration, and they may not all be available on a particular network. Some of the facilities provide an enabling function, so that individual calls can request particular values or facilities, using the facilities request field of the Call Request packet. This is described in the next section.

**Requesting facilities in the Call Request packet**   The facility field in the Call Request packet consists of a number of individual facility requests, each of which has a fixed size and format. As long as the field is inspected and decoded from the start there is no confusion over its meaning. However, to avoid trying to interpret the Call User Data as facilities, there is a length field of the overall request.

   Each facility request consists of a facility code followed by one or more parameters. The facility code indicates both the facility class, and how many parameters there are. The facility code belongs to one of four classes, depending on how many bytes the request requires. The first two bits of the code indicate the class as follows:

| Class | Bit values of facility code byte | Number of bytes of parameters |
|-------|----------------------------------|-------------------------------|
| A | 0 0 x x x x x x | 1 |
| B | 0 1 x x x x x x | 2 |
| C | 1 0 x x x x x x | 3 |
| D | 1 1 x x x x x x | variable |

For class D facility codes, the byte following the facility indicates how many further bytes there are.

   A network may offer facilities not defined in X.25, and these are requested by inserting a Facility Marker after any X.25 requests and before the non-standard requests. The Facility Marker is a class A request with all bits of the facility code set to zero.

   The X.25 facilities are encoded in the Call Request as shown in the following sections. (The use of the facilities was explained earlier in this section.)

**Closed User Group**   A DTE may belong to several Closed User Groups which are identified by a two-digit decimal number assigned by the network administration. The DTE may also belong to several bilateral Closed User Groups which are identified by a four-digit decimal number assigned by the network administration. When initiating calls to other members, or to the single other member, of a Closed User Group, the number must be given in the facility field, as well as the network address of the called DTE being given.

Closed User Groups are encoded either as a class A facility followed by the two-digit number encoded in one byte, or, in the case of Bilateral groups, as a class B facility with the four-digit number encoded in two bytes. The encoding is a simple one of each decimal digit being encoded in four bits.

For example, a call within group 78 would have facilities encoded:

| | |
|---|---|
| 00000011 | C. U. G. facility |
| 01111000 | to group 78 |

A call within bilateral group 7654 would have facilities encoded:

| | |
|---|---|
| 01000001 | Bilateral C. U. G. facility |
| 01110110 | |
| 01010100 | to group 7654 |

**Reverse Charging**   This is a class A facility encoded in one of the following two ways:

| | |
|---|---|
| 00000001 | Reverse Charging facility |
| xx000000 | Reverse Charging NOT requested |
| 00000001 | Reverse Charging facility |
| xx000001 | Reverse Charging requested |

The same facility code is used for requesting Fast Select, as shown below; thus there has to be an encoding of the second byte for not requesting use of the facility.

**Fast Select**   This is a class A facility encoded with a facility code of 00000001 as for reverse charging. The first two bits of the second byte are encoded as follows:

| | |
|---|---|
| 00xxxxxx | Fast Select NOT requested |
| 01xxxxxx | Fast Select NOT requested |
| 10xxxxxx | Unrestricted response Fast Select requested |
| 11xxxxxx | Restricted response Fast Select requested |

**RPOA selection**   This is a class B facility encoded with a facility code of 01000100. The two parameter bytes contain the Data Network Identification Code (DNIC) of the requested Recognized Private Operating Agency. The DNIC is covered in Chapter 7.

**Flow control negotiation**   The flow control parameters are packet size and window size, which may be requested in Call Request packets if the facility is subscribed to. The Call Accept also contains a facility field which may negotiate a different set of values for the parameters, within the limits described above in the explanation of the facility.

Packet size is a class B facility where the two parameter bytes are encoded with the values requested in the DCE-to-DTE and the DTE-to-DCE directions in that order.

The encodings are as follows:

| Value | | Packet size |
|---|---|---|
| 4 | (0100) | 16 |
| 5 | (0101) | 32 |
| 6 | (0110) | 64 |
| 7 | (0111) | 128 |
| 8 | (1000) | 256 |
| 9 | (1001) | 512 |
| 10 | (1010) | 1024 |

For example the DTE might request a packet size of 128 outgoing and 256 incoming. This would be encoded as

| | |
|---|---|
| 01000010 | Packet size facilities |
| 00001000 | 256 incoming |
| 00000111 | 128 outgoing |

The Call Accept packet in response to this might indicate packet sizes of 128 in both directions, perhaps because of limitations imposed by the network administration. The facilities in the Call Accept would thus be encoded as

01000010
00000111
00000111

The calling DTE would either work to these values or transmit a Clear Request.

Window size is a class B facility. The two parameter bytes indicate the values requested in the DCE-to-DTE and the DTE-to-DCE directions in that order. The values will be less than 8 in normal working and less than 128 if the extended numbering facility is being used.

For example the DTE might request a window size of 3 outgoing and 7 incoming. This would be encoded as

| | |
|---|---|
| 01000011 | Window size facility |
| 00000111 | 7 incoming |
| 00000011 | 3 incoming |

The Call Accept packet in response to this might indicate window sizes of 3 in both directions, perhaps because of limitations imposed by the network administration. The facilities in the Call Accept would thus be encoded as

    01000011
    00000011
    00000011

**Throughput class negotiation**   This is a class A facility with facility code 00000010. The parameter carries the class requested in the DCE-to-DTE and DTE-to-DCE directions in that order. These are encoded as follows:

| *Encoding* | | *Throughput class* (bit/sec) |
|---|---|---|
| 0 | (0000) | − |
| 1 | (0001) | − |
| 2 | (0010) | − |
| 3 | (0011) | 75 |
| 4 | (0100) | 150 |
| 5 | (0101) | 300 |
| 6 | (0110) | 600 |
| 7 | (0111) | 1200 |
| 8 | (1000) | 2400 |
| 9 | (1001) | 4800 |
| 10 | (1010) | 9600 |
| 11 | (1011) | 19 200 |
| 12 | (1100) | 48 000 |
| 13 | (1101) | − |
| 14 | (1110) | − |
| 15 | (1111) | − |

For example, the DTE might request an outgoing class of 2400 bit/sec and an incoming class of 300 bit/sec. The facility in the Call Request would be encoded as follows:

| | |
|---|---|
| 00000010 | Facility code |
| 01011000 | 2400 outgoing, 300 incoming |

**Call setup procedures**   The contents of the Call packet have been discussed in previous sections, and it has been shown that the Call Request is transmitted from one DTE to another. The X.25 recommendation differentiates between the Call Request transmitted from the calling DTE to the local DCE, and the packet transmitted from the remote DCE to the remote DTE. The latter is referred to as the Incoming Call packet.

This naming strategy reinforces the following points:

● The network need not be implemented using X.25 − and indeed usually is not − so the packets may not exist in the network in the same form or with the same significance as outside of it.

● The two DTEs may be able to have different facilities negotiated with the network. Thus what goes into the network may not be exactly the same as what comes out at the other end. For example, one DTE has a packet size of 256 and the other DTE has a packet size of 128. A full Data packet sent from the first DTE would have to be converted into two Data packets to be transmitted to the second DTE. Thus, in general, there is a relationship but not an equality between packets sent by one DTE and received by the other.

Note also that in general the LCN used by one DTE will be different to that used by the other DTE for the same call. This was explained in Chapter 1.

Figure 4.16 shows the correspondence between packet names used at either end of the call.

**Fig. 4.16** Correspondence between packet names at each end of the network

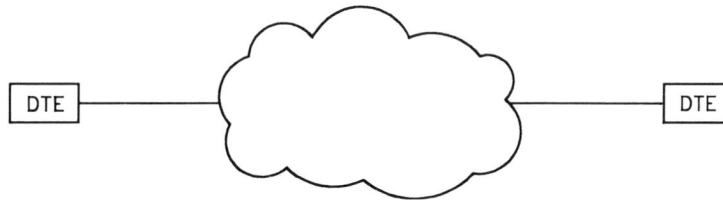

| Call Request | ⟶ | ⟶ | Incoming Call |
| Call Connected | ⟵ | ⟵ | Call Accepted |
| Clear request | ⟶ | ⟶ | Clear Indication |
| Clear Confirmation | ⟵ | ⟵ | Clear Confirmation |
| | | | |
| Reset Request | ⟶ | ⟶ | Reset Indication |
| Reset Confirmation | ⟵ | ⟵ | Reset Confirmation |
| | | | |
| Restart Request | ⟶ | ⟶ | Restart Indication |
| Restart Confirmation | ⟵ | ⟵ | Restart Confirmation |

**The M-bit**   The Data packet contains a bit called the M-bit, the "more" bit, or the "more data mark". If the M-bit is set, then it indicates that the data contained in the packet is continued in the next Data packet. Any number of Data packets may be connected in this way, and are refered to as a *complete packet sequence*. This is illustrated in Fig. 4.17. Note that the last packet of the sequence has the M-bit reset.

The reason for the sending DTE indicating a complete packet sequence is that it allows the application to transfer logical units of data

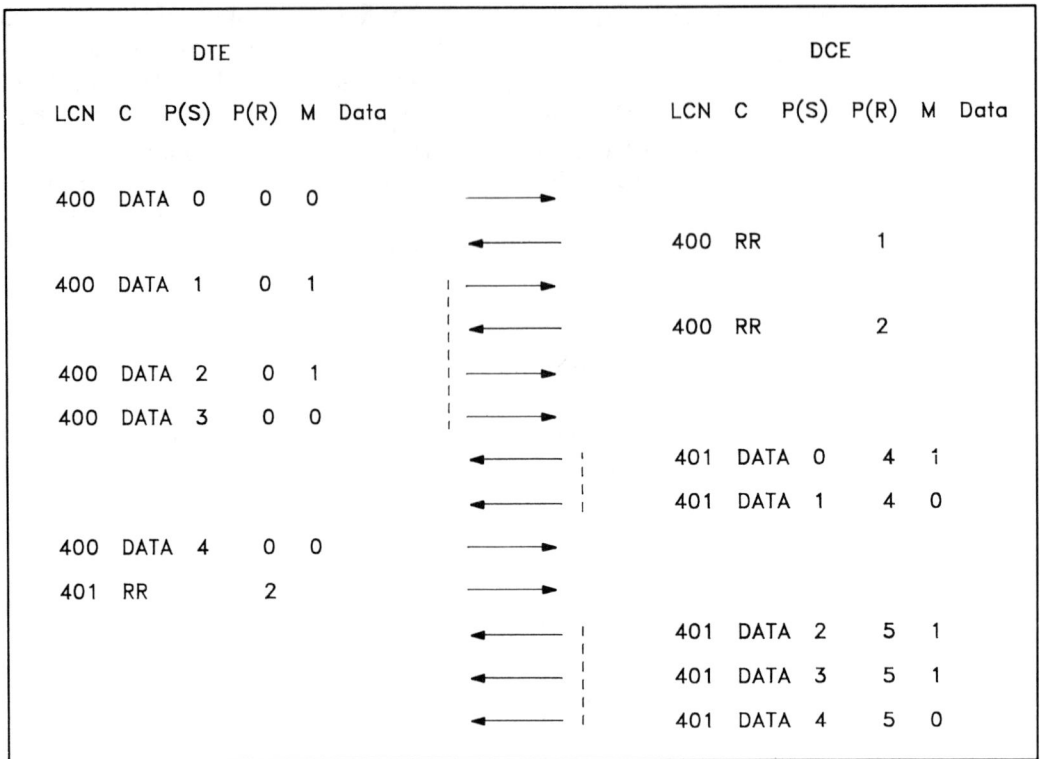

DTE

| LCN | C | P(S) | P(R) | M | Data |
|-----|------|------|------|---|------|
| 400 | DATA | 0 | 0 | 0 | |
| 400 | DATA | 1 | 0 | 1 | |
| 400 | DATA | 2 | 0 | 1 | |
| 400 | DATA | 3 | 0 | 0 | |
| 400 | DATA | 4 | 0 | 0 | |
| 401 | RR | | 2 | | |

DCE

| LCN | C | P(S) | P(R) | M | Data |
|-----|------|------|------|---|------|
| 400 | RR | | 1 | | |
| 400 | RR | | 2 | | |
| 401 | DATA | 0 | 4 | 1 | |
| 401 | DATA | 1 | 4 | 0 | |
| 401 | DATA | 2 | 5 | 1 | |
| 401 | DATA | 3 | 5 | 1 | |
| 401 | DATA | 4 | 5 | 0 | |

**Fig. 4.17** Use of the M-bit to indicate a complete packet sequence; the sequences are marked with a line

(blocks of a file perhaps) independently of the restrictions of the network. The network can then combine the data into a different number of packets prior to delivery to the receiving DTE. This will be an advantage if the receiving DTE has a larger packet size than the sending DTE, since the number of packets and therefore the protocol overhead is reduced. However, packets will only be combined if they are full. Packets that are not full will never be combined even if the M-bit is set.

Packets delivered to the receiving DTE following combination will have the M-bit set on all packets except the last in the new sequence. If the receiving DTE has a lower packet size than the sending DTE, then packets may need to be split by the network before delivery. Such packets will be delivered as a complete packet sequence with the M-bit set on all packets except the last.

Packet combination is likely to be an important issue in host-to-host traffic where full packets are commonly used. Typical PAD transmissions will not use full packets owing to the use of forwarding conditions. Data from a host to a PAD may be sent as a complete packet sequence and gain advantages from combination.

Note that packet combination may not occur if the D-bit is set (see below).

**The D-bit** The Data packet contains a bit called the D-bit or the Delivery Confirmation bit. This allows a DTE to verify that a particular packet has been received by the remote DTE. Normally, the acknowledgement sent by the DCE to a Data packet indicates that the packet has been received by the DCE, but does not give any further information.

If the D-bit is set on a particular Data packet from the DTE, then the DCE will only acknowledge it when the data has been delivered to the remote DTE. Thus the acknowledgement, by RR, RNR, or piggybacking, will be delayed while the data is transmitted across the network and the acknowledgement returns. At the remote end the data will be delivered by the DCE in a Data packet with the D-bit on. The remote DTE must acknowledge it as soon as possible so that the acknowledgement can be passed back through the network and the window re-opened at the transmitting DTE. Following transmission of a Data packet with the D-bit set, further Data packets may be sent, with or without the D-bit set, until the window is closed.

If a complete packet sequence is indicated using the M-bit, then the packets will not be combined if the D-bit is set. If a sequence contains a packet with the D-bit set, then packet combination will stop with this packet, and the last packet of the combined sequence will be marked with a D-bit. This is illustrated in Fig. 4.18 where the receiving DTE has twice the packet size of the transmitting DTE.

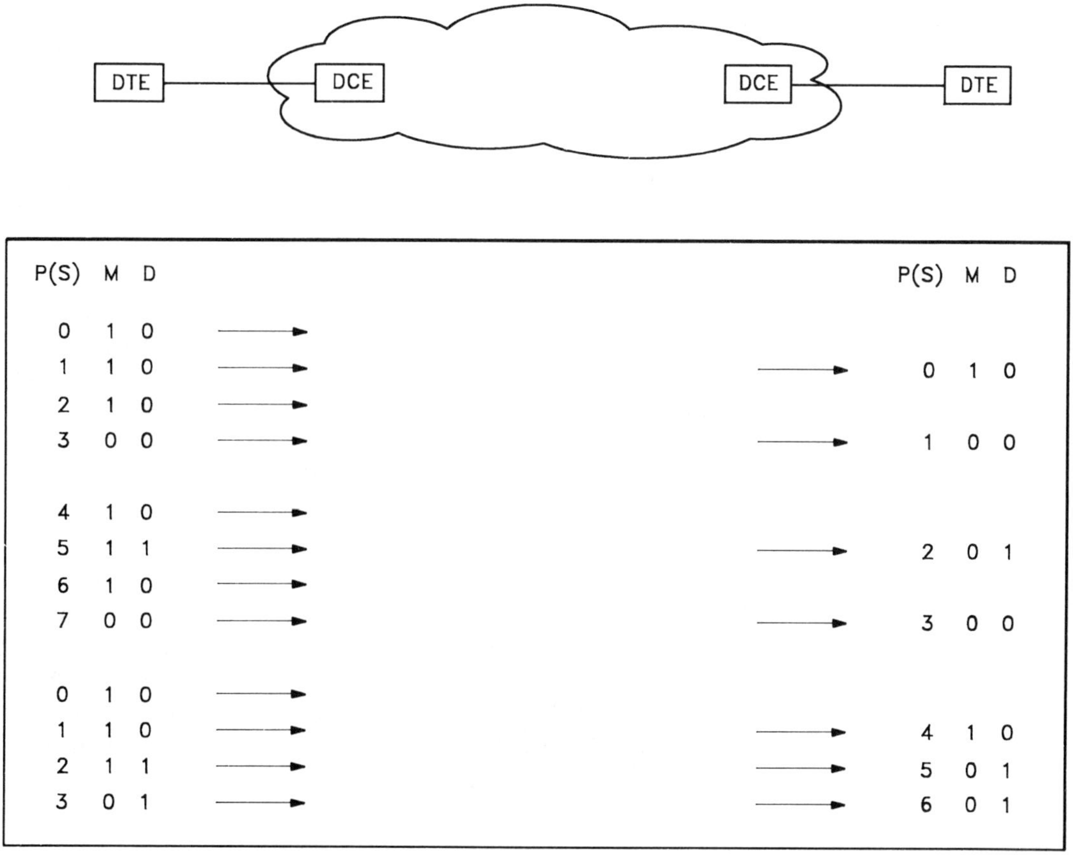

**Fig. 4.18** Effect of the D-bit and M-bit used together; the receiving DTE has twice the packet size of the sending DTE, and the packets are full

If a packet with the D-bit set has to be split owing to a smaller packet size at the receiving DTE, then the network will set the D-bit of the final packet of the sequence. The network will only acknowledge the original packet to the sending DTE when the receiving DTE has acknowledged receipt of the final packet of the sequence.

**The Q-bit**    The Q-bit or Qualifier bit is not used by the network, and is simply available for use between the two DTEs for their own purposes. It is commonly used to indicate Data packets that have a controlling function in the application. An example of this is in X.29, where the Q-bit indicates X.29 control packets. This is explained in Chapter 2.

**Interrupt packets**    The normal exchange of Data packets between two DTEs may not be adequate to cope with all situations that may arise. One example of this is where a terminal user wants to stop transmissions from a host computer. Sending an instruction to stop will have a delayed effect, owing to the time taken for the instruction to traverse the network, and owing to the amount of data that may already have been transmitted by the host and which is queued in the network awaiting delivery to the screen.

X.25 provides the *Interrupt* packet (INT) for these situations. The Interrupt is simply an indication sent from one DTE to the other, which is not subject to the flow control mechanism, and which is treated with higher priority than data packets by the network. It would be usual for an Interrupt packet to "overtake" Data packets in the network already sent by the DTE. It is therefore the quickest way for one DTE to send an indication to the other.

The meaning of the Interrupt is not defined in the X.25 recommendation, but it is conventionally regarded as a request for the remote end to stop what it is doing and take notice of instructions from the local end. The Interrupt contains a single byte of User Data that the local DTE can send to the remote DTE to qualify the interrupt.

The remote DTE responds to the interrupt packet with an *Interrupt Confirmation* packet (INC) which does not contain any User Data. A confirmation is then returned by the network to the DTE that sent the Interrupt. When the DTE receives the confirmation the interrupt procedure is complete.

Only one interrupt procedure can be in progress for a particular call and DTE at any time. This is because there is no means of relating confirmations to interrupts. The restriction allows the Interrupt and Confirmation packets to have a simple structure and therefore to be small, and the transmission can therefore be fast. Figure 4.19 shows an example of two DTEs exchanging data and interrupts. Note that the DTEs can have interrupt procedures in progress simultaneously, but that each can have only one per call at a time.

| LCN | C | P(S) | P(R) | Data | | LCN | C | P(S) | P(R) | Data |
|---|---|---|---|---|---|---|---|---|---|---|
| | DTE | | | | | | DTE | | | |
| 400 | DATA | 0 | 0 | | → | 400 | DATA | 0 | 0 | |
| 400 | DATA | 1 | 0 | | | 400 | INT | | | |
| 400 | DATA | 2 | 0 | | | 400 | DATA | 1 | 0 | |
| 400 | INT | | | | | 400 | INC | | | |
| 400 | DATA | 3 | 0 | | | 400 | DATA | 0 | 2 | |
| 400 | INC | | | | | 400 | DATA | 2 | 0 | |
| 400 | DATA | 0 | 2 | | | 400 | DATA | 3 | 0 | |
| 400 | DATA | 4 | 1 | | | 400 | INT | | | |
| 400 | INT | | | | | 400 | INT | | | |
| 400 | INT | | | | | 400 | DATA | 4 | 1 | |
| 400 | INC | | | | → | 400 | INC | | | |
| 400 | INC | | | | ← | 400 | INC | | | |

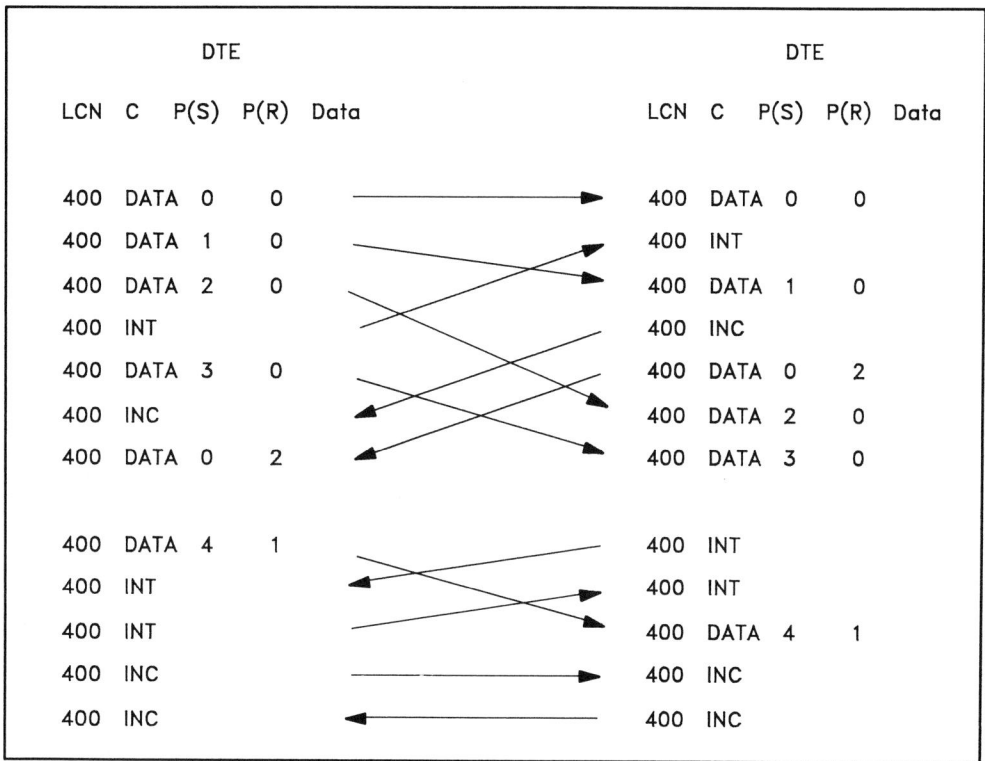

**Fig. 4.19** The Interrupt procedure; interrupts may "overtake" data already transmitted

**Reset packets** There may be occasions when things "go wrong" and a fresh start is required. Examples of such occasions are:

- An error is detected in the protocol; for instance a DTE transmits a second Interrupt packet to the DCE before the first has been confirmed.
- The DTE may run out of space and be unable to process packets correctly.
- The network may suffer congestion and be unable to process packets correctly.

A full list is given in Appendix B.

X.25 provides the Reset procedure for these occasions. This procedure causes both the network and the two DTEs to abandon any Data or Interrupt packets that have been received – whether or not they have been acknowledged – and to re-initialize the sequence number mechanism.

The procedure may be initiated either by the network or by one of the DTEs. When initiated by the network, both DTEs are sent a Reset Indication packet and must both return a Reset Confirmation to their respective DCEs. A DTE initiates the procedure by sending a *Reset Request* (RST) to the DCE, which is transmitted by the remote DCE as a *Reset Indication* to the remote DTE. The remote DTE responds with a

Reset Confirmation (RSC) which is passed back through the network to the local DTE to complete the procedure. Once the procedure is completed then all data and interrupts have been lost and the call is in the same state as if a Call Request and a Call Accept have been exchanged.

The Reset Indication packet contains two fields to inform the DTE why the Reset has been issued. The first of these is the Resetting Cause which is a single byte encoded as follows:

0   The Reset was requested by the remote DTE.
3   Remote procedure error – the network has detected a protocol violation at the remote end.
5   Local procedure error – the network has detected a protocol violation at this end.
7   Network congestion – the network is congested and cannot process data correctly.

Note that further encodings are defined in the recommendations but have been excluded here for clarity.

The second field is called the Diagnostic Code. It may be set by the initiator of the Reset to give further details of the reason – perhaps the

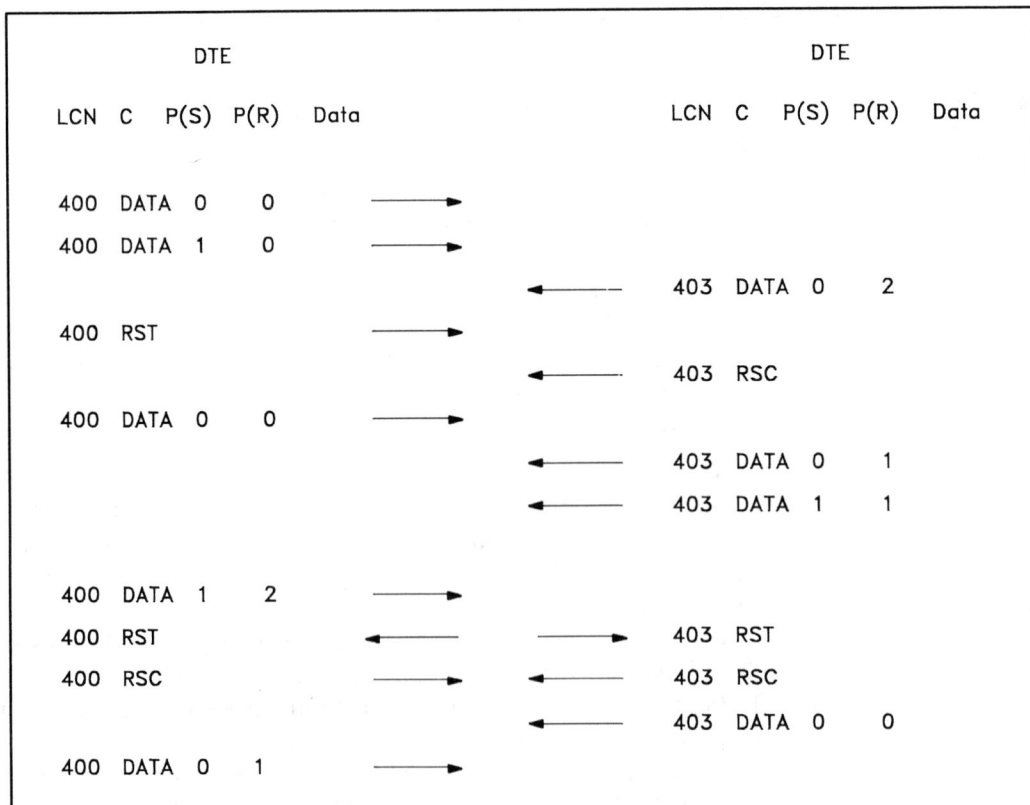

**Fig. 4.20** The Reset procedure; the first Reset is initiated by one of the DTEs, the second by the network and is transmitted to both DTEs

| DTE | | | | | | DTE | | | | |
|---|---|---|---|---|---|---|---|---|---|---|
| LCN | C | P(S) | P(R) | Data | | LCN | C | P(S) | P(R) | Data |
| 400 | DATA | 0 | 0 | | → | | | | | |
| 400 | DATA | 1 | 0 | | → | | | | | |
| | | | | | ← | 403 | DATA | 0 | | 2 |
| 400 | RST | | | | → | | | | | |
| | | | | | ← | 403 | RSC | | | |
| 400 | DATA | 0 | 0 | | → | | | | | |
| | | | | | ← | 403 | DATA | 0 | | 1 |
| | | | | | ← | 403 | DATA | 1 | | 1 |
| 400 | DATA | 1 | 2 | | → | | | | | |
| 400 | RST | | | | ← → | 403 | RST | | | |
| 400 | RSC | | | | → ← | 403 | RSC | | | |
| | | | | | ← | 403 | DATA | 0 | | 0 |
| 400 | DATA | 0 | 1 | | → | | | | | |

precise type of protocol error – and if not set explicitly will be delivered as zero by the network.

Figure 4.20 shows an example of two DTEs exchanging data and undergoing Resets. Note that different LCNs are in use at the two DTEs and that this will generally be the case. If a DTE and the local DCE transmit a RST to each other on the same LCN, because both want to Reset, then the procedure at that end is complete and no confirmations at that end are issued.

**Call clearing procedures**    The clearing procedure may be initiated either by the network or by one of the DTEs. When initiated by the network, both DTEs are sent a Clear Indication and must both return a Clear Confirmation to their respective DCEs. A DTE initiates the procedure by sending a *Clear Request* (CLR) to the DCE, which is transmitted by the remote DCE as a *Clear Indication* to the remote DTE. The remote DTE responds with a *Clear Confirmation* (CLC) and the network then delivers a CLC to the local DTE to complete the procedure. When the procedure is completed then all resources used by the call are released for use by other calls. In particular, the logical channels are de-allocated and may be used for new calls.

The Clear Indication contains two fields to inform the DTE why the Clear has been issued. The first of these is the Clearing Cause field which is a single byte encoded as follows:

| Hex | Decimal | Reason |
|-----|---------|--------|
| 00 | 0 | The Clear was requested by the remote DTE. |
| 01 | 1 | Number busy – all available incoming and bothway LCNs at the remote DTE are in use. |
| 09 | 9 | Out of order – the link between the remote DCE and DTE is not operational. |
| 11 | 17 | Remote Procedure error – the network has detected a protocol violation at the remote end. |
| 19 | 25 | Reverse Charging Acceptance not subscribed – the Call Request included a facility requesting reverse charging, but the remote DTE does not subscribe to the Reverse Charging Acceptance facility. |
| 29 | 41 | Fast Select Acceptance not subscribed – as above but for Fast Select facility. |
| 03 | 3 | Invalid facility request – the Call Request contained a facility request that was invalid in some way. For example, it may request the Reverse Charging facility, but the DTE does not subscribe to the ability to initiate these calls. |
| 13 | 19 | Local Procedure error – the network has detected a protocol violation at this end. |
| 05 | 5 | Network congestion – the network is congested and cannot process packets correctly. |

Note that further encodings are defined in the X.25 recommendation but have been excluded here for clarity.

Many of these clearing causes are the same as may be given for Resets or Restarts, for instance any of the three types of packet may result from a network congestion. The type of packet sent in any given situation will depend to some extent on the severity of the conditions, and to some extent on the decisions made by individual implementors.

The second field is called the Diagnostic Code. It may be set by the initiator of the Clear procedure to give further details of the reason — perhaps the precise nature of a protocol violation — and if not set explicitly will be delivered as zero by the network.

If a DTE and its local DCE transmit a Clear Request and a Clear Indication to each other on the same LCN, because both want to clear the call, then the procedure at that end is complete and no confirmations at that end are issued.

The Clear Request is not subject to the flow control mechanism of Data packets, and in general will overtake Data packets queued in the network. This is illustrated in Fig. 4.21.

**Fig. 4.21** The Clear procedure; the Clear Request may "overtake" data already transmitted and thus clear the call and cause the data to be discarded

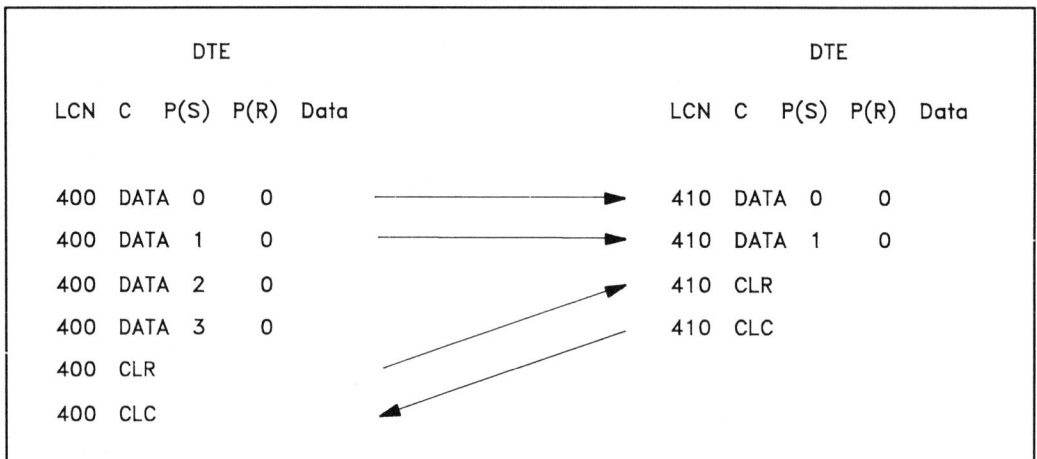

| DTE | | | | | | DTE | | | | |
|-----|---|------|------|------|---|-----|---|------|------|------|
| LCN | C | P(S) | P(R) | Data | | LCN | C | P(S) | P(R) | Data |
| 400 | DATA | 0 | 0 | | | 410 | DATA | 0 | 0 | |
| 400 | DATA | 1 | 0 | | | 410 | DATA | 1 | 0 | |
| 400 | DATA | 2 | 0 | | | 410 | CLR | | | |
| 400 | DATA | 3 | 0 | | | 410 | CLC | | | |
| 400 | CLR | | | | | | | | | |
| 400 | CLC | | | | | | | | | |

The network will usually discard Data packets and Interrupts for the virtual circuit once the Clear procedure has started, thus, in general, sending a Clear Request will cause loss of data. This situation can be remedied by the use of higher-level protocols. An example of such a remedy is the Invitation to Clear message of X.29 which is described in Chapter 2.

### 4.3.5 Permanent virtual circuits

A Permanent Virtual Circuit (PVC) is a circuit permanently assigned between two DTEs. As long as all equipment involved in the circuit is operational then data can be exchanged between the two DTEs. No Call is possible nor is a Clear. The circuit is permanently in the Data transfer state.

Data, Interrupts, flow control, and Reset packets may be exchanged as for Switched Virtual Circuits. The Reset takes on more significance with PVCs because it is used to indicate the state of the circuit.

The Reset packet has encodings defined for PVCs to indicate that the circuit is out of order, and that the Remote DTE – and by implication the whole circuit – is operational.

A Restart operating on the DTE/DCE interface will Reset any PVCs.

### 4.3.6 Timers and numbers

The timers used with the packet layer of X.25 are as follows (there are no numbers such as N1 as for the link layer). There are two sets of timers: the first set applies to the DCE if the DTE should fail to respond.

- T10, *Restart timer*  Following transmission of a Restart Indication, the DCE waits this time to receive a Restart Confirmation or a Restart Request.
- T11, *Call Request timer*  Following transmission of an Incoming call, the DCE will wait for this time to receive a Call Request, a Call Accept, or a Clear Request.
- T12, *Reset Request timer*  Following transmission of a Reset Indication, the DCE will wait for this time to receive a Reset Confirmation or a Reset Request.
- T13, *Clear Request timer*  Following transmission of a Clear Indication, the DCE will wait for this time to receive a Clear Confirmation or a Clear Request.

The second set of timers applies to the DTE should the DCE fail to respond:

- T20, *Restart Request timer*  Following transmission of a Restart Request, the DTE will wait for this time to receive a Restart Confirmation or a Restart Indication.
- T21, *Call Request timer*  Following transmission of a Call Request, the DTE will wait for this time to receive a Call Connected, a Clear Indication or an Incoming call.
- T22, *Reset Request timer*  Following transmission of a Reset Request, the DTE will wait for this time to receive a Reset Confirmation or a Reset Indication.
- T23, *Clear Request timer*  Following transmission of a Clear Request, the DTE will wait for this time to receive a Clear Confirmation or a Clear Indication.

If any of the timers expire then recovery action may be taken to try and complete the procedure. The action will depend on the timeout that expires, but may be the sending of a Clear Request or Indication, or a Diagnostic Packet. Some administrations may allow retransmission of some types of packet for which a response is required.

The X.25 recommendation defines the following values for the DCE timers:

$$T10 = 60 \text{ secs}$$
$$T11 = 180 \text{ secs}$$
$$T12 = 60 \text{ secs}$$
$$T13 = 60 \text{ secs}$$

The recommendation defines the following limits for DTE timers:

T20 max of 180 secs
T21 max of 200 secs
T22 max of 180 secs
T23 max of 180 secs

## 4.4    Differences between X.25 (1980) and X.25 (1984)

The newer version of X.25 makes a number of changes which are explained in this section. The fundamentals of X.25 are unchanged, and the new features are generally additions to the recommendation. Some interworking is possible between the two versions, and many manufacturers offer products conforming to a subset of the 1984 standard.

### 4.4.1    Starting the link

As well as the SARM and SABM frames for starting the link a new frame type called SABME is available. This extended SABM is used on links which have subscribed to extended operation — that is they have frame counters working to modulo-128. The use of extended operation is still an option chosen when the user subscribes to the link to the network. Thus, in general, either SABM or SABME will always be used unless the user arranges a different set of facilities with the network administration. The arrangement of new facilities may be performed on-line as explained later, so it is possible to use both SABM and SABME on a given link.

### 4.4.2    Multilink procedure (MLP)

MLP allows a layer two link to be distributed over several physical circuits, thus increasing the reliability and/or the throughput of the link. This is a subscription option chosen by the user when the link to the network is obtained. The frame format is slightly more complex than shown for the 1980 links, and has sequence numbers working to modulo-4096.

Each of the individual links is started by a normal SABM or SABME, and those that start successfully are available for use by the MLP. The MLP is initialized by the exchange of MLP Reset and Confirmation frames using any of the available links. Data can be exchanged using any of the links, and a particular MLP frame may be transmitted on several links to increase the chance of successful delivery. The receiving MLP will reassemble the data in the correct sequence and pass it to layer three for processing.

### 4.4.3  Interrupt packets

Interrupts may now contain up to 32 bytes of user data instead of just one. The Interrupt confirmation packet is unchanged.

### 4.4.4  Cause fields in Reset and Restart Requests

In the 1980 recommendation a Reset or a Restart issued by a DTE had to have a cause of zero. In the 1984 recommendation the code may be zero, or may also be any value with the high-order bit set. The interpretation of the values is a matter between the two DTEs.

### 4.4.5  On-line facilities registration

A new packet type has been defined, the Registration Request, which allows the DTE to request changes to the current facilities agreed with the DCE. The ability to send such requests is itself a facility agreed between the user and the network administration.

The DCE responds to all requests by transmitting a Registration Confirmation packet, which indicates the new current values of all facilities. The number of facilities that the DTE may request to change is dependent on whether or not calls are currently in progress over the link. If calls are not in progress then the following may be requested:

D-bit modification (slightly different use of the D-bit procedure, used
    for DTEs implemented prior to the current use)
Packet retransmission
Extended packet sequence numbering
Change of logical channel ranges for "incoming only"
Change of logical channel ranges for "outgoing only"
Change of logical channel ranges for bothway calls

The following facilities may be requested at any time:

Charging information (see below)
Throughput class negotiation
Flow control parameter negotiation
Reverse Charging acceptance

Fast Select acceptance
Outgoing calls barred
Incoming calls barred

Values may be negotiated for the following:

Throughput class
Non-standard default packet sizes
Non-standard default window sizes


### 4.4.6  New facilities

The following new facilities are available. Each must be agreed at subscription time when the user obtains the link to the network.

*Local charging prevention*
   This facility prevents charges being incurred by the local DTE. Thus:

   ● Reverse charge calls to the local DTE will not be transmitted by the DCE.
   ● Outgoing calls must all request Reverse Charging, or indicate where the charges are to be made by using the Network User Identification facility (see below).

*Network User Identification*
   This facility allows the DTE to indicate the identity of other DTEs for purposes such as charging. Thus, a Call Request may use this facility to send the charges for the call to another user. The format, use and restrictions of this facility are defined by the network administration.

*Charging information*
   This facility may either be permanently present if agreed between the user and the network administration, or be selected as a facility for individual calls. When present, the DCE will send information about the call charges to the DTE that will have to pay these charges. The information is sent when the call is cleared in either the Clear Indication or Clear Confirmation.

*Hunt Group*
   This facility is similar to having several telephone lines into a building, all sharing the same telephone number. When the facility is agreed, the user is assigned a number of lines into the network, each formed of a DCE-DTE link, but with a single Hunt Group address. Calls to the user specify the Hunt Group address in the Call packet, and the network will assign the incoming call to any free logical channel in the Hunt Group.
   The DTEs forming the Hunt Group may have specific addresses as well as the Hunt Group address, so a caller can specify precisely where the call is to be connected. In this case the address returned to the

caller in the Call Accept or Call Clear will be the specific DTE address.

If the caller specifies the Hunt Group address, then the address returned may be either the Hunt Group address or the specific DTE address. If the address is different to that requested by the caller, then this is indicated by the Called Line Address Modified Notification facility (see below).

*Call Redirection*

This facility allows a Call Request to a particular DTE to be redirected by the network to another specified DTE. The redirection is activated if the original called DTE cannot be reached, owing to its being out of order, or if all of its incoming logical channels are in use.

Depending on the network, the process may be limited to a single redirection, or may use multiple redirections in order to connect the call. In the latter case the network may either allow a list of alternative DTEs to be associated with the DTE subscribing to the facility, or allow the process to chain through DTEs each with a single associated alternative.

The called DTE that eventually receives the Call Request may be informed of the redirection by the network. This is achieved by the network setting the Call Redirection facility in the Incoming Call packet. When redirection takes place the original calling DTE is informed of the redirection by the network. This is achieved by the network setting the Called Line Address Modified Notification facility (see below).

*Called Line Address Modified Notification*

This facility is used by the network to inform a calling DTE that a Call Request has been passed to a different called DTE to that specified in the request. The facility is set by the DCE in the Call Connected or Clear Indication returned to the DTE. The packet will contain the address of the actual DTE that received the call. The address may be modified either because of Call Redirection or because of a call to a Hunt Group. The reason is encoded in the facility.

*Call Redirection Notification*

This facility is used by the network to inform a called DTE that the Call Request is a result of redirection. The facility is set by the DCE in the Incoming Call packet which will contain an indication of whether the original called DTE was out of order or busy. The packet will also contain the address of the original called DTE.

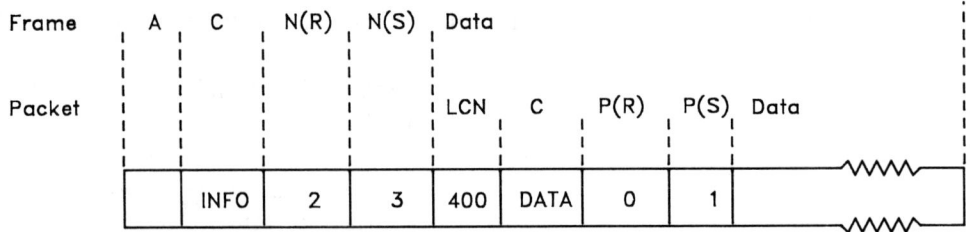

Fig. 5.1 Conceptual
packet format

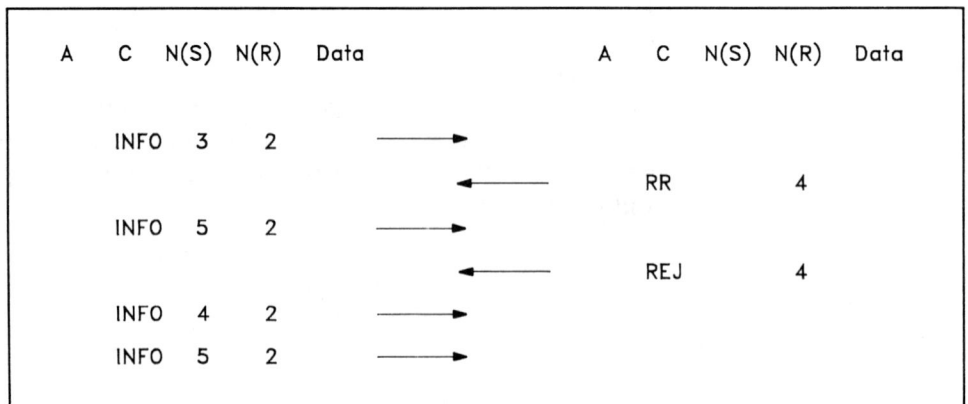

Fig. 5.2 Frame exchange

# 5 Network management

## 5.1 Introduction

This chapter looks at the problems of the network manager in monitoring and controlling the equipment making up the network. This is partially concerned with the ubiquitous Network Management System which many manufacturers hold up as the answer to all network problems. But, without an understanding of what is actually happening on the wires and in the boxes, the Management System is unlikely to provide the manager with the power to resolve problems in the network.

To be effective, the network manager should not just react to faults when they occur. It is much more important to evaluate continually the network configuration, to look continually at what is happening at a very detailed level, in order to prevent problems developing into failures.

## 5.2 Information from the protocols

Looking at the packets and frames being transmitted on a link is probably the most effective way of determining the health of the connection. This is because in a single view, information on the cable, the modems, the network components, and the applications software is obtained.

We have already looked at the basic method of transferring data in Chapter 1, and this is summarized in Figs. 5.1 and 5.2.

The frame is the layer two unit of transmission and is exchanged between two topologically adjacent components. The Info frame carries layer three packets, and errors are detected by the use of the FCS and by the send and receive sequence numbers. Info frames may be acknowledged by RR frames to open the transmission window when there is no returning data. The Reject frame causes retransmission of a frame that has not arrived, and all subsequent frames.

The packet is the layer three unit of transmission and is exchanged between two end-points of the network, typically a host service and a user of the service. It carries an LCN allowing several calls to share a layer two link, and packet level send and receive sequence numbers.

With this set of frames and packets – which is just a small subset of the available types – we can already gain useful information about what is happening on a link.

- Info frames always carry layer three packets so the proportion of other frame types indicates the layer two protocol overhead. For instance, if 50% of the frames were not Info frames then there is a great deal of layer two protocol, which has implications on how much layer three data can be carried.
- A similar argument applies to data packets at layer three.
- The number of flag bytes compared to the number of frames indicates the unused capacity of the line. If there are more flag bytes than frames in a particular direction then the transmitter is not keeping the line busy, either because it has nothing to send or because it has to delay whilst the frame is created.
- RR frames are a little difficult to interpret since they may be sent for a number of reasons:
  1) To open the window if Info frames are uni-directional for a period and the window has closed.
  2) To acknowledge frames as they are received.
  3) They may be sent after a period of inactivity to acknowledge all the frames received so far. This has the advantage of keeping the window open even if data is uni-directional.
  4) They may be sent even if no frames are awaiting acknowledgement in an "idle handshake" to inform the other end that this end is still here.

  Acknowledgement of each and every frame is usually wasteful since in most conversations the data is sufficiently bi-directional to allow piggybacking of the RR on returning Info frames. Data is also often delivered in bursts, a screenful of data for instance, and the flow of Info frames should be fast enough not to need each one acknowledging individually.
- A similar variety of reasons may exist for the use of an RR packet at layer three.
- The Reject frame indicates that data has arrived out of sequence. This is rarely due to the sender getting "mixed up" or "forgetting" to send a frame, and rarely due to the cable or modems "dropping out" and failing to deliver a frame. It is almost always due to a frame being corrupted during transmission, delivery or reception, and the FCS check therefore failing. This check is usually performed in the hardware, and if it fails the corrupt frame is simply discarded so that layer two never receives it.

Suppose that there is a simple link from a PAD into a public network, and that the PAD has accumulated the figures shown in Fig. 5.3. What do these show us?

The first thing we must know is over what period the figures were accumulated. Otherwise it is usually impossible to see if this is a busy

**Fig. 5.3** Example
statistics

```
┌─────────────────────────────────────────────────────────────┐
│                                                               │
│   Layer 2     INFO      400  out     360  in                  │
│                                                               │
│               RR        300  out     350  in                  │
│                                                               │
│               REJ         1  out       2  in                  │
│                                                               │
│                                                               │
│   Layer 3     DATA       50  out     300  in                  │
│                                                               │
│               RR        300  out      50  in                  │
│                                                               │
└─────────────────────────────────────────────────────────────┘
```

five minutes or a very light month. Suppose in this case it is over one hour. This shows immediately that the link is lightly loaded, since the capacity of a 2400 bps link is $2400/8 \times 3600$, i.e. over 1 Mbyte in the hour. If the PAD had reported the number of flag bytes sent and received then it would have been a huge number, and for this reason it is actually rarely used.

The PAD sent 400 Info frames and got 350 RR frames in acknowledgement. This shows that the frames were probably fairly far apart and most required individual acknowledgement. The network is not polling RR frames informing the PAD that the network is operational, otherwise there would be many more.

There were a total of three Reject frames. Whether this is high or low depends on the quality of the circuit and therefore the expectations that it is reasonable to have. A good circuit would be characterized by the following features:

● Good-quality cable used for all connections, probably featuring braided screen, thick conductors, and of a length allowed by the appropriate electrical standard. These features are discussed in more detail in Chapter 7.
● Modems that are able to distinguish real signal from induced noise.
● Wiring practices that avoid potential problems – this is discussed in Chapter 7.

On a good-quality circuit there should be no Rejects; on the other hand, one or two a day is not enough to cause any worry.

It is much easier to create a poor-quality circuit by performing any selection of the following:

● Using cheap cable not soldered properly.
● Running data cables close to mains cables and equipment.
● Using equipment with faulty components.
● Mis-configuring equipment.
● Using voice grade telephone circuits.

How much this affects the protocol is an open-ended question. Links exist where there is a Reject frame for every half dozen correct frames, and the users are probably unaware of it because the correction is automatic. In such cases it is difficult to say that the error rate is unacceptable even though it is very bad. Like a crackling telephone line it is up to the users to decide what is desirable and acceptable.

At layer three in the example, each data packet has been acknowledged by an RR. This the normal rate of acknowledgement. Comparing the number of Data packets with the number of Info frames, we can see that coming from the network most of the Info frames carry real data (300 out of 360); or layer three RRs (50 out of 360). This leaves ten Info frames carrying other types of packet.

Data travelling from the PAD has resulted in 50 packets and 300 RRs. In this case however there are 50 additional packets carried by layer two Info frames. What these packets might be was shown in Chapter 4; however, they represent a significant number compared to the number of Data packets (50 compared to 300) and it can be seen that there is cause for worry and further detailed investigation.

## 5.3 More examples of statistics

Figure 5.4 shows an example of some statistics. These figures show a lack of any information transfer but a large number of frames out of the unit. This is due to the unit transmitting DM frames to try and establish the link. Such transmissions are normal since it is what the protocol requires, but it is fairly evident that the other end is not going to respond. In this case it may be sensible to stop the operation of the port by a command to the software, and the processor resources can then be used more effectively.

**Fig. 5.4** Example statistics

| Layer 2 | INFO | 0 out | 0 in |
|---------|------|-------|------|
| | RR | 0 out | 0 in |
| | REJ | 0 out | 0 in |
| | Total | 78000 out | 0 in |

Figure 5.5 shows some higher numbers. Considering layer three first, the number of Data packets out is much the same as the number of RR packets in, and thus the packets are separate and individually acknowledged. Only around half of the incoming Data packets are acknowledged, so these are probably closer together.

The layer three totals indicate a few packets not yet accounted for. These may be due to the following packet types: Call, Call Accept, Clear, Clear Confirm, and Reset and Restart procedures.

The number of Info frames both incoming and outgoing is slightly greater than the total number of packets. This is due to errors in transmission, which is confirmed by the presence of Reject frames.

The layer two RRs indicate that many of the frames were acknowledged by piggybacking, or else were sent sufficiently quickly to not

Fig. 5.5 Example
statistics

| Layer 2 | INFO | 78197 | out | 123550 | in |
|---------|------|-------|-----|--------|-----|
| | RR | 88019 | out | 34001 | in |
| | RNR | 0 | out | 0 | in |
| | REJ | 23 | out | 28 | in |
| | Total | 166243 | out | 157586 | in |
| Layer 3 | DATA | 22979 | out | 101298 | in |
| | RR | 55108 | out | 22005 | in |
| | INT | 0 | out | 1 | in |
| | Total | 78172 | out | 123517 | in |

require individual acknowledgement. The layer two totals indicate a few frames not accounted for, which are due to link setup procedures.

Figure 5.6 shows some more statistics. The layer three Data packets transmitted by the component are almost all acknowledged, whereas packets received by the components are only acknowledged at the rate of about three to one. This indicates that outgoing data is relatively sparse, and incoming data is, in general, continuous.

Fig. 5.6 Example
statistics

| Layer 2 | INFO | 16330 | out | 23512 | in |
|---------|------|-------|-----|--------|-----|
| | RR | 19476 | out | 30816 | in |
| | RNR | 0 | out | 0 | in |
| | REJ | 0 | out | 4 | in |
| | Total | 35815 | out | 54337 | in |
| Layer 3 | DATA | 12319 | out | 11310 | in |
| | RR | 4001 | out | 12127 | in |
| | Total | 16330 | out | 23506 | in |

The layer three window size is two by default. Unless a different default has been subscribed to, or most calls negotiate a different number, then many of the incoming Data packets must be acknowledged by piggybacked RRs on outgoing data. If this were not the case then the three-to-one ratio of RRs to data would not be possible.

The layer three totals show a very small number of packets unaccounted for. These are due to resets and restarts, and call setups and clears.

At layer two the number of Info frames is practically equal to the total number of layer three packets. The difference is due to errors in transmissions and subsequent retransmission of the frames. This is confirmed by the small number of layer two Rejects.

The number of RRs received is greater than the number of Info frames transmitted. This must be due to the other end of the link sending RRs to indicate that it is operational. These cannot have been polled since there is no corresponding number of RRs transmitted.

The total number of frames indicates very few frames unaccounted for. These are probably due to the exchange of frames during Disconnected Mode. As they are so few, the link must have been established soon after this component became operational and started collecting statistics.

## 5.4 Collecting the statistics

Statistics of the type shown in the previous sections are available in practically every network component, though manufacturers naturally differ in the details of collection and presentation. The statistics are made available through the operator interface, and various commands are usually available to see the statistics for individual parts or for the whole unit.

Figure 5.7 shows an example of the statistics that may be collected for the whole unit. This shows the time for which the unit has been running, from which it could be inferred whether the unit had ever restarted. It also gives information on the time over which the statistics have been collected.

**Fig. 5.7** Example of overall statistics for a network component

| | |
|---|---|
| Running time: | 27: 15: 41 |
| Incoming call requests: | 137  (110 accepted) |
| Outgoing call requests: | 47   (47 accepted) |
| Currently active calls: | 18 |
| Highest number of active calls: | 18 |
| Total packet count: | 717864 in   820799 out |
| Memory usage: | 90% highest  72% average |

Information on the highest number of simultaneous calls handled by the unit gives a rough indication of traffic densities within the network, though this must be read in conjunction with the packet totals. Memory usage is very useful in determining whether the unit is overloaded, and may confirm information gained from the number of RNRs.

## 5.5 Remote collection

The nature of a network is that it is spread over an area. In the case of an X.25 network, the area may be international. It is clearly not generally possible for the network manager to visit all sites and obtain statistics, therefore some means of accessing the data remotely is required.

Most manufacturers provide a simple X.29 host-end module in their equipment, which allows a user on a triple-X PAD to call into the equipment and access internal services. These services may be as follows, and will vary from manufacturer to manufacturer:

- Services which generate text, allowing the functionality of the network components to be verified.
- Configuration, allowing the unit to be configured remotely, for example to reduce the range of LCNs supported on a link to give a better service to fewer users.
- Statistics, allowing access to the history of the unit.

Using these services, the network manager can remain at his desk once all the components have been installed, and can monitor the network and tune it.

## 5.6 Running the network

Operation of a large network can be a tremendous task involving many different functions. Some of these are shown in the following sections. However, it must always be remembered that network management is about keeping the network running, not explaining why it has stopped.

### 5.6.1 Component inventory

This is a list of all the components in the network, showing for each one things like: serial number; manufacturer; date of purchase; location (building, room number); maintenance contractor; history (last maintenance, faults).

The list is used for keeping track of the equipment – knowing what is faulty, what has gone away to be mended, and what is temporarily in its place. It can also give valuable information on trends, such as "all PADs from manufacturer A fail with a CPU fault after one year in service".

### 5.6.2 Connectivity

This is a diagram of how the network is connected together. It shows for example that port seven of the switch at site A is connected to port 3 of the switch at site B. At a further level of detail it shows that the link is accomplished using modems and a circuit provided by a public authority. It then links to the component inventory to show the precise items.

The diagram is valuable when a fault develops since the manager knows which ports on which units to look at, and can use the statistics creatively. For example, if a user telephones the network manager and complains that a terminal has gone dead, then the manager can do the following:

- Look up on the diagram which PAD the user is connected to. It may be worth looking in the component inventory to see if the PAD or user terminal has a history of problems.
- Call into the internal services of the PAD.
- If the call is successful then the configuration of the users port can be examined, and perhaps some other information gained from the PAD. It is probably possible to determine what the PAD thinks is happening on that port.
- If the call is not successful then it appears that the PAD has stopped. The manager could then call the internal service of an adjacent unit and determine the state of the connecting link. It may be that the PAD is operational but one link has gone down and this was the one that both the user and the manager were trying to use.
- The manager could then change the configurations in the PAD and adjacent units, using the diagram, so that all calls were routed a different way.

This type of operation would be difficult without a diagram, since it is unlikely that the manager is able to remember which port connects to which throughout the network. Even more dangerous is the probability of getting it wrong, and affecting calls that are working.

### 5.6.3  Line inventory

Some information on the physical circuits forming the network is necessary, in order to be able to investigate faults in them. For lines installed within a building then probably just the route of the cable is recorded. For circuits provided by public authorities then the circuit number, its termination points and information about the authority are needed. There will often be a fault-reporting procedure for leased lines, which should be recorded in the line inventory.

Continuing the example of the previous section where a link fault was detected, then the network manager may go through the following procedure:

- Look at the statistics for the ports at each end of the link to determine whether problems have been occurring for some time — perhaps a large number of REJ frames — or whether the loss was sudden. There may be a record in the connectivity log of a few rejects being seen last week, indicating that the problem has been building up.

- From the component inventory the manager can determine whether or not the PAD, switch and modems support V.54 loopback testing. This is a means of remotely testing a circuit and was explained in Chapter 3.
- The manager then tries a local component loop, local analog loop, and remote digital loop test from each end of the link by calling into the relevant internal services.
- If both sets of tests fail on the remote digital loopback, then it is clear that the fault lies in the actual circuit, and this should be reported to the public authority using the procedure in the line inventory. There may be a note in the inventory stating that, when a loopback test was tried last week, some errors were found and the authority advised of impending problems. This sort of information clearly lends more weight to the manager's assertion of where the blame lies.

Had the line been a purely local one then it is likely that the manager will have to visit both ends, and perhaps walk the length of the wire, to find the fault. At least with an inventory the manager knows where to walk!

## 5.7  Managing the network

There is a difference between running the network and managing it. In the example of the previous section it was fairly easy to respond to the incident by calling adjacent components, isolating the fault, and re-routing calls. However, it would have been much better if the calls had been re-routed the previous week when the potential for an incident was discovered, and for test calls to have been left running to bring on the full failure.

The person who never opens the bonnet of the car should not be surprised when it breaks down. The person who sees that more oil is being used than is normal, or that there is a seepage from a cooling hose, can get to a garage.

Network managers should never be seen idle. They should always be making calls around their network, getting information, and writing it down in the various logs. It is only by this proactive information gathering that subtle changes can be seen, faults predicted, and full user service maintained despite the faults.

So what should the network manager do? The answer is everything. At least the following:

- Look regularly at all statistics in all components and analyze in detail what the figures say about what is happening. What do they say about the quality of transmission? What do they say about how the link is being used? Is there too much traffic? are the various X.25 implementations compatible?

● Schedule times when loopback tests can be performed to get more information.

● Write everything down so differences can be seen and trends analyzed.

## 5.8   Network management systems

A Network Management System (NMS) is a computer system that automates some of the tasks of the network manager, leaving the manager free to use insight and intuition.

Something that nearly every NMS can do is to produce a diagram of the network on a screen, and colour the diagram to indicate operational and non-operational components. Such a diagram is useful in showing a working network, and perhaps in showing a general area of failure, but as previous sections have shown it cannot perform the prediction of faults that is so much part of a manager's function.

Network management systems work in one of two basic ways which are characterized by the way in which the data is obtained from the network components. This is either by a *terminal call* or by a *special protocol*.

### 5.8.1   Terminal access

An NMS of this type obtains data from network components by making a call in the same way as a human operator would, using the internal services of the components. The NMS contains software which emulates a terminal call, and has to be configured with the commands to send to the components, and the format of the returning data. Such a system is not too efficient since it has to overcome the verbose and unformatted interface designed for people. However, as long as the NMS can hold different configurations, it can manage components from different manufacturers simultaneously.

The network manager can use this system to pursue investigations by giving commands directly to the NMS. The NMS then passes the information request to the relevant component using the dialect that the component needs. The network manager then has a unified approach to problem solving, and can use the technology to aid the investigations.

Perhaps the single greatest disadvantage with this approach is that it relies on calling into the network at the worst possible time − when a failure has occurred and the network is suffering a crisis. Examples of this would be:

● Investigating a unit where the problem was lack of memory. The extra memory usage to handle the NMS call may exceed the capabilities of the unit, and cause it to restart some or all of the lines.

- Investigating a line where the problem is faulty hardware. The extra traffic generated by the call may cause more errors to occur, which may rise to a level at which every packet is corrupted.
- Investigating a failed unit where all the calls are being re-attempted on different routes. The congestion caused by this may prevent the NMS call getting through.

A further consequence of this type of NMS is that for the display to be up-to-date, the components of the network have to be called often. On a large network this means that a considerable proportion of the network traffic may be due to NMS calls being switched around the components to their final destination.

### 5.8.2 Proprietary protocols

An alternative method of operating the network management system is to use a special protocol between the components and the central system. Such a protocol is designed for efficiency, which means that there can be a permanent connection to the NMS. This permanent connection allows the components to send unsolicited messages to the NMS about events that are occurring, and the problems of polling at the worst time are avoided.

This approach is the subject of investigative work by ISO, and is likely to be standardized in a way that allows different manufacturers equipment to cooperate in the same managed domain. Until the standardization, network managers have to use a proprietary protocol and restrict themselves to components from one source.

### 5.8.3 The Camtec network management system

One such proprietary protocol is that designed by Camtec Electronics Ltd., a UK company that produces a range of networking equipment. The Camtec system is presented here as an example of the type of benefits that can accrue from the use of a full protocol, and other systems may offer similar or better benefits. My grateful thanks must go to Camtec for their permission to reproduce this information. It is emphasized that the description given here does not form a definition of the product, nor does it cover all the features of the product. The Camtec system is subject to the normal development process, so many additional capabilities may be present in the system currently available. Further details, including the latest product description, can be obtained direct from Camtec by writing to the address shown at the end of the book.

**The protocol** Each network component is preconfigured with an address table specifiying an X.25 port on the component, and an address on the network. When the software in the component starts running, then the network management module attempts a call to the specified address on the specified port to connect to the central network management system. If the call attempt is not successful then it is re-attempted at increasingly longer intervals. This establishes the connection between the component and the NMS, and an initial exchange of messages identifies the unit.

If each component were to have a permanent individual call then there would be a considerable impact on the network. Minimal impact is one of the claimed advantages of this type of protocol, and normally the components are configured to share a virtual circuit. This is illustrated in Fig. 5.8.

**Fig. 5.8** Network components sharing virtual circuit to NMS

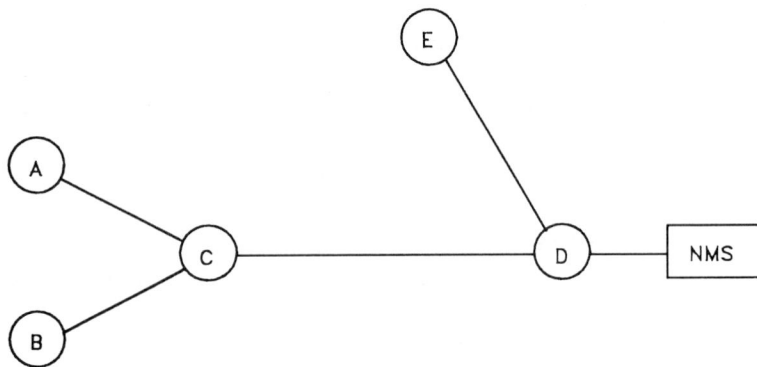

When components A and B start running then they attempt their NMS calls using the links to component C. Components C and E attempt their NMS calls using the links to component D, and this has a direct link to the NMS.

The preferred way of powering-up the network is component D first, then E and C, then A and B. When D starts, it establishes a call to the NMS and then exchanges protocol messages to start a dialogue. These protocol messages are over and above X.25, and simply use X.25 as a carrying service.

When component E starts then a call is established to D using a virtual circuit in the normal way. The routing tables in D cause the call to be accepted at this point, but the Network Management Protocol messages are routed over the existing virtual circuit to the NMS. This is shown in Fig. 5.9.

The figure shows the software modules used in component D. Module Y controls the X.25 call to the NMS, and this passes protocol messages from module Z which handles all network management activity for component D. The X.25 call from E is handled by module X, which again passes the protocol messages to module Y for onward routing to the NMS. The protocol messages all contain an identification of the component to which they refer, so there is no confusion at the NMS.

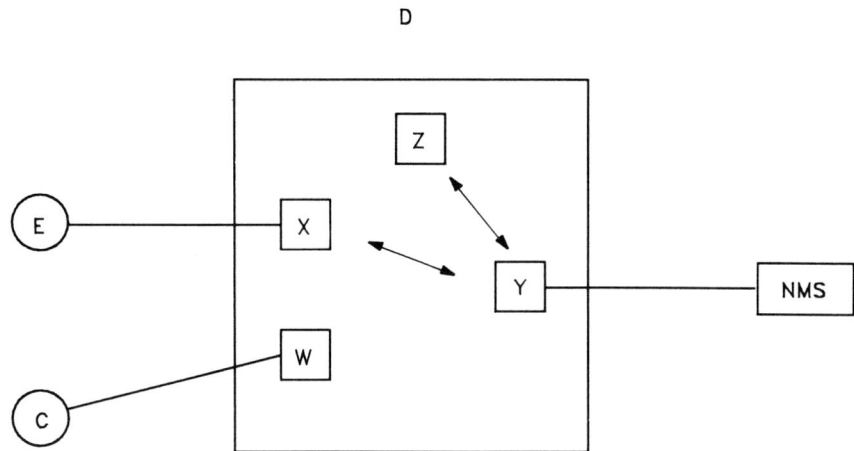

**Fig. 5.9** Multiplexing of network management data within a component

Likewise, messages from the NMS identify where they are going, so software module Y can route them correctly.

When component C starts then its X.25 call is handled by module W, and again the messages are passed to and from module Y for multiplexing to the NMS.

The same multiplexing process occurs with components A and B at C, so in fact the messages between W and Y are themselves multiplexed for three components. This multiplexing process ensures that a minimal number of virtual circuits are used in the network, and since the calls are permanently running, there is only an overhead for setting the call going when a component starts to run.

Notice that once the X.25 call from D to the NMS is established, then there is no further X.25 overhead on the wire − other than data − whatever happens in the network. The single call carries all the management traffic.

The protocol also allows multiple calls into the NMS, thus sharing the management traffic over a number of calls. Using the same network, the routing in component D could be arranged as shown in Fig. 5.10. In this case the call from C is routed as a normal switched X.25 call onto the link to the NMS. No attempt is made in component D to decode the data. There are therefore two virtual circuits on the NMS link that each carry data for components in the network: one for A, B and C, and one for D and E.

Using similar techniques the network management traffic can be shared over several physical circuits.

**The network management system**   The Camtec NMS is a computer, linked to the network by X.25. As such, the system can be accessed using devices locally connected to the computer, or it can be called from any point in the network as any other host would be. Such network calls use standard triple-X protocols and do not involve the actual management calls.

**Fig. 5.10** Mixed
multiplexing and direct
calls for network
management data

D

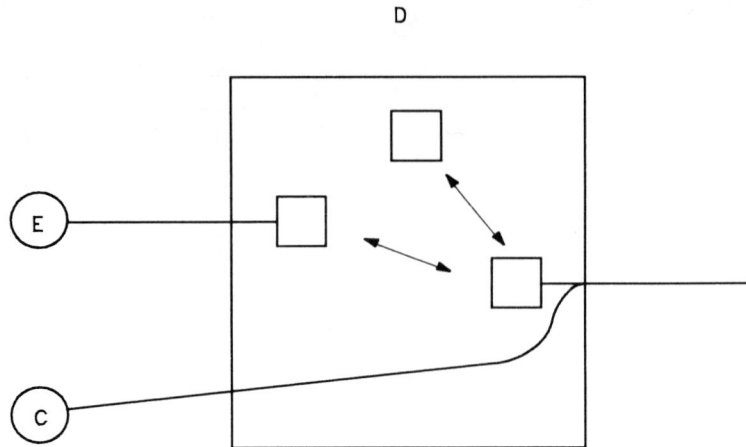

The NMS allows several different classes of user, and the class is associated with the logon identification of the user. This allows a network manager to access the system from any point in the network, and to retain the full privileges enjoyed when using a terminal directly attached to the NMS.

The normal management of the network is carried out by users classed as operators. When an operator logs on to the NMS, a menu giving access to eight sub-systems is presented, all of which are related by a database at the heart of the NMS.

The following sections show the features of these sub-systems.

**Testing sub-system**  The testing sub-system allows the operator to request components in the network to perform loopback tests, to determine the quality of modems and physical circuits. The components use the V.54 controls described in Chapter 3.

The operator is able to specify which loopback test to perform, and the duration of the test. The results can be examined when the test is complete, and partial results can be examined as the test is running. The results are stored in the database, allowing the operator to monitor degradation in performance at leisure.

**Control sub-system**  The control sub-system gives access to the individual physical ports of each component in the network, and allows the operator to start, stop, or close activity on the port. These terms were described in Chapter 2.

The operator is also able to control the Network Management Protocol on a link, without affecting other calls. This may be useful if a component that was multiplexing a number of calls goes down and has to be replaced. The Network Management Calls then need to be re-established. The sub-system also allows the operator to restart units where more gentle techniques of problem solving have not been successful.

**Monitoring sub-system** This is one of the most valuable sub-systems and provides access to all the statistics held by network components. Two groups of figures can be requested: status, which is what is happening at this instant; and statistics, which is a summary of what has happened over time. The sub-system allows the operator to reset the statistics to zero, so the time over which collection occurs can be controlled.

Each group of figures can be requested for the whole component, for a single port, or for every port. The figures for the component are items such as the time that the component has been running and the total number of calls and packets that have been handled.

For ports, the following information is returned. This allows an analysis of the health of the port to be performed as shown at the beginning of this chapter.

**Status:** what is happening at this instant
*Asynchronous lines*
State of significant modem signals
Logical PAD state
Whether logged-on
Whether a call is up
Flow control state

*X.25 lines*
State of significant modem signals
Whether the X.25 link is up
Whether any calls are active over the link
The logical layer two state
Last frame type sent and received

**Statistics:** the accumulated history of the unit
*Asynchronous lines*
Current call duration
Number of segments in and out for current call
Number of "breaks"
Number of resets
Number of line framing errors
Total number of calls
Total durations of all calls
Total number of segments in and out
Summary of clearing code types

*X.25 lines*
Total number of call attempts in and out
Total number of successful calls in and out
Number of current calls

Maximum number of simultaneous calls

Number of transmit and receive errors

Totals of frame types in and out for Info, RR, RNR, REJ, FRMR, total of all types

Totals of packet types in and out for Data, RR, RNR, REJ, RES, INT, total of all types

Total number of data segments in and out

Summary of clearing code types

Number of packets of network management data

The data segments count is included to aid in the analysis of charges, either incurred by the network from public bearers, or imposed by the network administration on its users. In either case, charges are normally made per segment. The definition of a segment varies somewhat, but is basically as follows:

- One for every data packet up to 64 bytes
- One for every extra 64 bytes in a data packet
- One for other types of packet such as call, reset or interrupt.

Transmit and Receive errors are detected by the hardware, and are the result of an overload of the software within the network components. The system of transmission and reception is explained below, and whilst it is common to many manufacturers, it must be emphasized that this is not general for all components from all manufacturers.

Between frames, the hardware sends flag bytes automatically. As soon as the processor requests frame transmission, then the hardware stops sending flags and instead sends the data that it receives from the processor. When the processor indicates the end of the frame then the hardware calculates the FCS, sends the FCS, and then returns to sending flags. A transmit underrun error occurs if the processor fails to deliver data to the hardware fast enough for the speed of the line. The hardware is compelled to transmit something, so the frame is corrupted.

This results in a problem. When the software fails to deliver data, the hardware may assume that the end of the frame has been reached, and thus calculate a correct FCS and send a correct frame. There is no way for this error to be detected unless the frame format is incorrect. In the case of an Info frame carrying a Data packet, the underrun may occur in the user data portion, and the data is therefore corrupted with no possibility of detection by the protocols at layer two or layer three. Fortunately the hardware alerts the software of the transmission of the FCS, and there is therefore a second chance for the software to prevent the error. Since the FCS is already being transmitted when the alert arrives, there is no way of sending the frame correctly; however, the software can instruct the hardware to perform an abort. The abort is a sequence of seven or more contiguous one bits that is defined in the X.25 recommendation for this type of situation. Since this number of one bits contravenes the normal rules for frame construction, it is easy for the receiving hardware to detect. It then informs the receiving software of the abort so that the frame can be discarded.

It should be emphasized that the responsibility for responding to transmit underruns and preventing undetectable errors lies with the software. It must be written in such a way that it can always respond to the hardware fast enough to initiate the abort sequence before the end of the frame.

On reception, the hardware automatically discards flag bytes. As soon as a frame starts then the hardware interrupts the processor, which must then take the data from the hardware at the rate they are received from the line. If the processor fails to take a character before the next is received, then the character is discarded by the hardware and the frame received by the processor is corrupt. This is a receive overrun error.

Manufacturers normally employ a small amount of buffering in the hardware which alleviates the processor response requirements, but does not remove the possibility of underruns and overruns.

Overruns and underruns indicate that the processor is unable to cope with the speed of the line, as well as perform the other tasks required of it. If these errors occur then the load on the processor must be reduced. Note that layer two will recover from these errors by retransmission.

Whenever statistics are requested, the results are stored in the database for later analysis or comparison with other figures. Apart from statistics gathering, information on the activity of a unit may be obtained from alarms. An alarm is generated by a unit in response to a preset condition, and results in a message being displayed on the operator screen. Optionally, the alarm may also result in the raising of a Trouble Ticket, which is an entry in the database. Whilst the operator may clear the alarm, the Trouble Ticket will remain as a reminder of the original situation causing the alarm. Alarms are explained in more detail in the next section.

**Configuration sub-system**  The configuration sub-system allows the operator to change the configuration of the network components. The same functions are available as if the operator were locally connected, and include such things as address tables, port configuration, text messages, and software patches.

One of the most useful items of configuration is alarms. An *alarm* is a message sent automatically by a network component to the NMS to inform it of an event that has occurred. This means that the NMS is informed of potentially serious events automatically, and does not have to poll the components to ensure that everything is alright.

Many alarms are based on an error rate. For example, if the number of retransmissions per hour on a particular port exceeds a given number, then an alarm is raised.

The following alarms typify those that the operator can configure:

X.25 line going down
Number of current calls
Transmission errors
Retransmissions
CPU usage
Line utilization

When an alarm is received by the NMS the information is stored in the database and, depending on the priority of the alarm, all operators have a message displayed on their screen. The priority is set by the operator when the alarm is configured.

An alarm can also be configured to raise a Trouble Ticket. This is a database entry which serves as an ongoing reminder of the situation causing the alarm. It therefore provides a tool to aid the management of faults that may occur on the network.

**Inventory sub-system**   The inventory sub-system is a direct interface to the database, allowing the operator to query data recorded by other sub-systems, or to access general information stored for each component. This sub-system allows the operator to record details such as serial number, location, service history, and maintenance contractor, thus avoiding the need for writing down the information. The use of a database means that the information need only be recorded once and cross-referred, so there is no repetition of information and it is easy to update.

**Reports sub-system**   Each NMS has a set of its own reports, which are an automatic way of querying the database and relating items of information together. For example, a report might show details of line error rate correlated against modem manufacturer. The reports are set up using a report writer package by a class of user called developers. Operators can only use the reports, not define them.

**Housekeeping sub-system**   The housekeeping sub-system is an interface between the operator and the underlying operating system of the NMS computer. The operating system is deliberately hidden in order to present a user-friendly system oriented to the needs of network management. The functions of this sub-system are as follows:

- Read and Send electronic mail to other users of the NMS. An operator might send mail to a developer for example, to request a new report.
- Find out who is using the NMS.
- Change the password of the user.

**Graphics sub-system**   This sub-system determines how the components of the network are connected together – from the information in the database – and draws a network diagram on the screen. The diagram is colour coded to show the particular parameters that the operator requests.

Various levels of detail can be obtained in the diagram, from an overall picture of the network, through progressively more detail, to individual units. This type of display depends on the network being arranged in a coherent, hierarchical manner. As long as this is the case, then the operator can "zoom-in" to a problem as follows.

At the highest level is a trunk display showing the entire network. This will appear as circles representing the regions of the network, connected by lines representing the trunk lines. Each of the circles in this display can be expanded to show more detail for the region. This will generally consist of a major switch providing access from the region to the trunk network, and the main switching PADs of the region.

The next display will show a sector of the region. This will show two of the switching PADs and all the connected PADs.

At the most detailed level, the individual network components can be shown. The display then shows the status of the connections to the unit. The individual links can also be displayed. This will show the line, the units that it connects, and the status of the connection.

At any level of the hierarchy, the operator can go up or down the display by choosing where to expand the display, or to reduce it. The operator can choose for the diagram to be colour coded to show status or statistics.

- If status is selected then colours indicate the operability of the equipment:

  *Green* (operational)
  Equipment operating normally.
  *Blue* (serviceable)
  The equipment is installed but has not yet been brought into service.
  *Orange* (reported)
  A fault has been discovered on the equipment and reported to the maintenance engineer.
  *Red* (faulty)
  The equipment is faulty and has not yet been reported.

- For X.25 lines the status display shows what the line is doing:

  *Green* (operational)
  The line is operational.
  *Magenta* (testing)
  A loopback test of the line is in progress
  *Yellow* (closing)
  The line is being taken out of service. No new activity is allowed but the current activity can continue.
  *Orange* (stopped)
  The line has been taken out of service.
  *Red* (down)
  The line is faulty.

- For asynchronous lines the display again shows what the line is doing:

  *Green* (operational)

      A call has been made from the terminal and the line is in data transfer state.

  *Blue* (idle)

      No call is in operation and the line is idle.

  *Magenta* (command)

      The line is in command state.

- When statistics are selected then the colours indicate the number of calls in progress on the item of equipment.

  | | |
  |---|---|
  | *Green* | No calls |
  | *Yellow* | 1—8 calls |
  | *Orange* | 9—20 calls |
  | *Mauve* | 21—50 calls |
  | *Red* | More than 50 calls |

Whilst the screen is showing a graphic display, then the operator can request regular polls to be performed automatically by the system to keep the screen display up to date. The poll requests the necessary data from the components involved in the display.

**Network manager facilities**   The operators of the network management system can perform all of the day-to-day functions of managing the network. In overall control of the NMS is the network manager, who performs the administration of the computer behind the management software.

Amongst the facilities available are the ability to assign and remove logon names, and the ability to set up a network poll. The network manager can also perform system backups, monitoring of system activities, and the initial work required to configure the graphic displays.

When assigning logon names and passwords the manager must also assign the class of the user, such as "operator", so that the relevant facilities are then available to the user. Apart from operators there are three other classes that can be given to users: helpers, developers, and installers.

The helper can use all of the features available to operators, but on an enquiry-only basis. The helper cannot change the contents of the database. This status is normally given to a small number of local advice staff, who can then investigate problems and answer queries from network users. This removes everything but difficult problems from the network operators.

The developer prepares report formats for use by the other classes of users. This is done with a report generator program, and the types and numbers of reports are therefore not predetermined by the system.

The final user status is installer, and is assigned to people who actually install the network components out on site. This is normally done as a two-stage process. The operator enters skeleton information to the database such as the component type, the site name, and the configuration of the component. The installer then completes the rest of the details after the unit has been installed, such as the room number, which port connects where, where the car park is, and who the site manager is.

The manager can configure the NMS to automatically poll the components in the network, to ensure that data in the database is current. Clearly, in a network of several hundred components the manager must be wary of polling all components too frequently, or the network will be swamped with management data and be unable to perform its function. The NMS therefore offers a number of options to balance the need for currency of information with impact on the network.

# 6 Going beyond X.25 – the seven-layer model

## 6.1 Introduction

Throughout this book the emphasis has been on people using networks to access computers. Whilst this type of traffic provides a convenient vehicle to discuss the network, it must be appreciated that nowadays large volumes of data may be sent directly from one machine to another. This host-to-host traffic is a significant feature of many modern networks.

Data sent in this way may use the capabilities of the underlying network in exactly the same way as the triple-X terminal protocols use them. An example of this is X.400, which is a recommendation describing how Electronic Mail messages are handled, and how the Message Handling System interfaces to the network. For this interface to be effective it is necessary for both the network and the application to fit into the same overall protocol strategy, which is where the Seven-Layer Model comes in.

The basic function and outline of the seven-layer model was discussed in Chapter 1. This chapter will expand the discussion to look at the recommendation in more detail, and will examine how the protocol stack works. In addition the place of some of the recommendations that are referred to most often will be discussed.

The model is described in ISO 7498, Open Systems Interconnection Basic Reference Model. The document provides a framework into which other standards and specifications may be placed in a coordinated way, to make systems open to each other. An open system is simply one which recognizes the framework of the model, and which adheres to some of the standards within the framework.

This chapter then goes on to look at some of the "wider" details of X.25, in particular how the international networks are created. Finally, the Systems Network Architecture (SNA) of IBM is introduced as another example of a layered communication system.

## 6.2 Layering

The functions necessary to achieve open systems interconnection are split into seven layers. The layers provide increasingly complex functions, so that any user of the open system can communicate with another. The model refers to the users as Application Processes − and people are called manual application processes.

At a conceptual level, a layer is referred to as the (N)-layer, and this is concerned only with the layer below, the (N − 1)-layer, and the layer above, the (N + 1)-layer. The (N)-layer has no knowledge of other layers within the open system. The (N)-layer provides a service to its user, the (N + 1)-layer, by using the services of the (N − 1)-layer.

The (N)-layer within the open system communicates with the (N)-layer in the open system to which it is connected, and this communication is carried out using the (N)-protocol and is shown in Fig. 6.1. The communication is achieved not horizontally, but vertically, using the services of the (N − 1)-layer. Thus, to send an (N)-protocol message from open system A to open system B, the message is first sent to the (N − 1)-layer, which has to communicate with the (N − 1)-layer in open system B. The (N)-layer message is (N − 1)-layer data. The whole process repeats at this new lower level, and the communication is achieved using the services of the (N − 1)-layer of the (N − 1)-layer as shown in Fig. 6.2.

The process is rather like a rich householder sending a message to a neighbour. The message is from householder A to householder B, but householder A will in fact give the message to the butler to send. Butler A will send the message to butler B for delivery to householder B but will ask the footman to take it. Footman A will send the message to footman B to give to butler B, but will ask the chambermaid to take it. Eventually the message will get to the gardener who will pass it over the fence to the other gardener.

All communication between the households − and between the communications stacks − is peer-to-peer. That is, at any level, the communication is between entities of a similar rank, and is achieved using the services of the lower layer. An example of this is that in an X.25 network all layer three protocol and data is carried as data in layer two Info frames.

Services of a layer, the (N)-services, are provided at the boundary between the (N)-layer and the (N + 1)-layer, and this boundary is the (N)-Service-Access-Point.

There is, of course, no (N + 1)-layer above layer seven, since the layer is used directly by the application process. Similarly, there is no (N − 1)-layer below layer one, since layer one is designed to use the physical medium directly.

Open systems that are in communication with each other need not be directly connected; there may be open systems between them providing a relay function. This is illustrated in Fig. 6.3 for an X.25 network.

Open system A is a PAD and the manual application process − the person − is in communication with an application in the host which is open system B. They are linked by C which is an X.25 switch. The lower

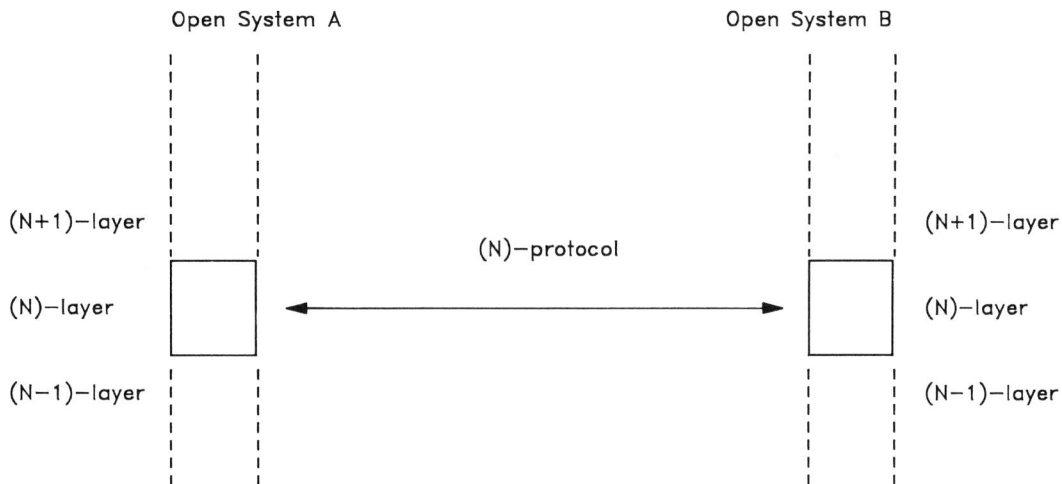

**Fig. 6.1** Peer-to-peer communication in layers of the model

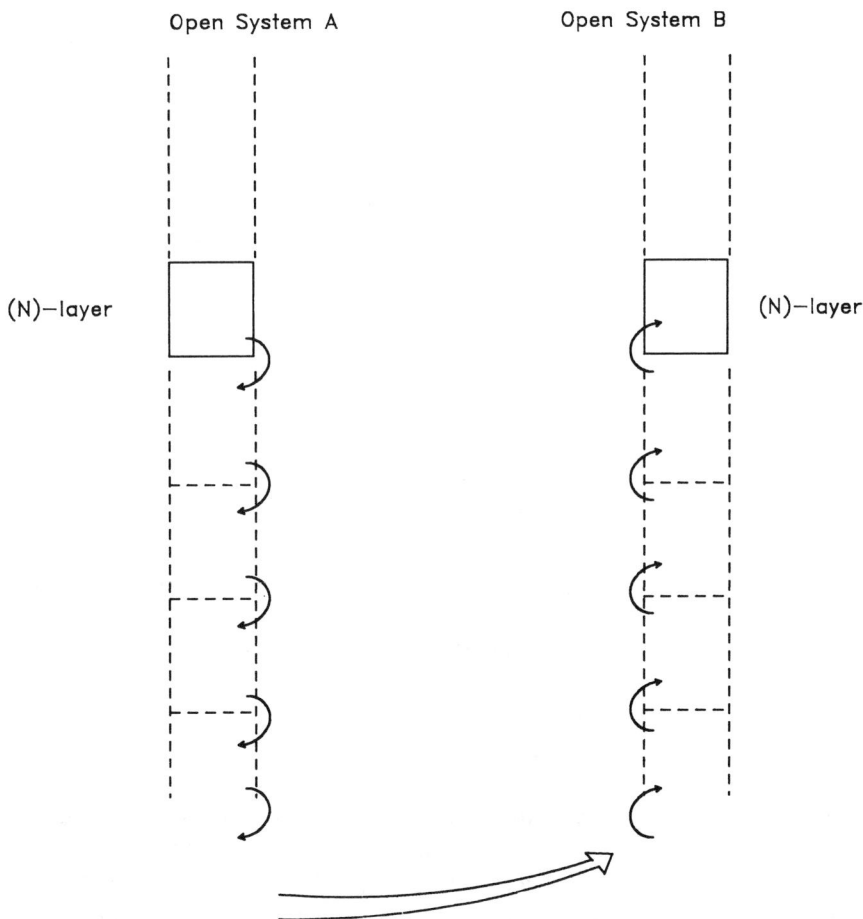

**Fig. 6.2** Actual communication is achieved by sending (N)-protocol messages using transmission services of the next layer down; the process repeats to the bottom of the stack, and reverses in the other open system

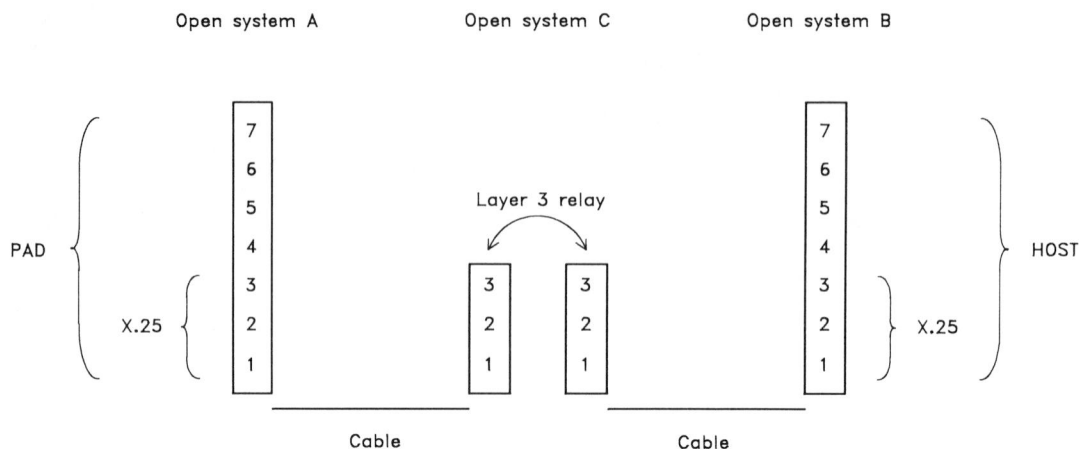

Open system A          Open system C          Open system B

```
        ┌─┐                              ┌─┐
        │7│                              │7│
        │6│                              │6│
        │5│         Layer 3 relay        │5│
PAD ⎰   │4│            ⌒                 │4│  ⎱ HOST
        │3│     ┌─┐       ┌─┐           │3│
  X.25⎰ │2│     │3│       │3│           │2│ ⎰X.25
        │1│     │2│       │2│           │1│
        └─┘     │1│       │1│           └─┘
                └─┘       └─┘
        ────────────      ────────────
           Cable              Cable
```

**Fig. 6.3** Use of a relay system in the lower layers

three layers of A and B send (N)-protocol messages to the relevant (N)-layer in the switch, whilst the upper four layers of A and B send (N)-protocol messages to each other without any knowledge of the intervening component.

The seven layers defined in the recommendation are as follows:

7 *Application layer*
6 *Presentation layer*
5 *Session layer*
4 *Transport layer*
3 *Network layer*
2 *Data Link layer*
1 *Physical layer*

### 6.2.1   Application layer

This is the window between the application process and the open system, and it enables the application to use the OSI environment. Two categories of service are commonly available depending on the type of the application process.

The first category provides a general or common interface for use by any application process, but which is not optimized for any particular application. These services are known as *Common Application Service Elements* (CASEs) and a typical one is Commitment, Concurrency, and Recovery (CCR). This provides for reliable continuation of an application process if a problem occurs.

Interfaces in the second category are provided on the assumption of a certain type of application process, and thus are tailored for the specific needs of that application. These are known as *Specific Application Service Elements* (SASEs). An example of this might be an application process to perform a File Transfer and Access Method (FTAM). A number of different applications exist that perform FTAM operations but all have similar needs of the open system. There is therefore an advantage in having an application layer tuned to these needs.

Similar applications in this second category are Job Transfer and Manipulation Protocols (JTMP) for remote submission of work packages to computers; Message Handling Protocols such as X.400; and Virtual Terminal Protocols (VTP).

### 6.2.2   Presentation layer

This layer is concerned with the way in which the application process data is represented. For example, the application may be an Automatic Teller Machine outside a bank, which provides a four-bit code indicating which button on the keypad has been presssed. The presentation layer may be required to represent these codes in an eight-bit form according to some pre-determined algorithm, in order that the lower layers can handle it. Likewise, when the eight-bit data arrives at the destination presentation layer, it may need to be represented as a sixteen-bit pattern before the computer can process it.

Each end of the presentation layer protocol only needs to know the requirements of its particular application process, and the manner of transfer within the open system. And the ends only need to agree on the transfer method between them.

The protocol of the presentation layer may be used to establish this agreement both at the start of a session and during the session to change the way that data is represented within the OSI environment. This may be needed for example when a computer changes from sending a textual data stream to sending graphic data to a terminal.

The definition of how data is structured is a complex subject, and to ensure commonality ISO have published a method of doing this. This is called Abstract Syntax Notation One (ASN. 1) and is published as ISO/DIS 8824.2. The use of ASN. 1 results in a rigorous definition of how the layer should transform the data that it handles.

Often the application processes will already use a data encoding which is suitable for transfer between the presentation entities. For instance both application processes may use ASCII, which may also be able to be handled by the lower layers of the protocol stack. In this case no transformation is needed.

### 6.2.3   Session layer

This layer allows for the orderly and controlled transfer of data between application entities. When the application is starting up, it will establish a session to the appropriate other application entity by giving a session address to the session layer and requesting a session connection. The two entities then have a connection and can exchange data.

In the event of a failure beneath the session layer, the session layer may have the capability of recovery without abandoning the session. This capability includes features to inform the application of a temporary pause, and to resynchronize activities.

When the application has finished, then the session layer is informed and the session is released. Within the session various parameters can be negotiated, including the manner of transfer, either two-way simultaneous or two-way alternate.

### 6.2.4  Transport layer

The transport layer is not only the mid-point of the model but is very much a half-way layer between those deeply concerned with the application and those deeply concerned with the mechanics of the network. Its function is to use the capabilities of the lower layers in order to provide the data transfer required by the upper layers.

The requirements are specified by parameters passed between the session layer and transport layer when a transport connection is established between two session entities. There are a number of parameters that can be negotiated at this time, but the key feature of the transport layer is that it provides quality of service. This means that it provides whatever speed, accuracy, and reliability are required by the upper layers, using the actual capabilities of the lower layers.

A second feature of the transport layer is that it will provide the required service as efficiently as possible.

Five classes of transport layer are defined:

Class 0  (*Simple*) – assumes a reliable network service. If an error occurs in lower layers then the session layer is informed and the transport connection is broken. There is no recovery.

Class 1  (*Basic error recovery*) – can recover from errors reported by lower layers. It is therefore often used with networks such as X.25, where errors are detected but not necessarily corrected.

Class 2  (*Multiplexing*) – allows several transport connections to share a network connection. This minimizes the cost of the connections whilst providing the required quality of service.

Class 3  (*Multiplexing and error recovery*) – combines the features of class 1 and class 2 to provide a very reliable and efficient transport service over a network that is basically reliable but which may report errors.

Class 4  (*Error detection and recovery*) – uses a complex system of timers and checks to detect errors, and can then recover from them. It is therefore used with networks that may lose or corrupt data without detecting the loss. A prime example of this is connectionless networks where data may be lost or arrive out of sequence without notification.

Where a requested quality of service cannot be met by a single network connection, then the transport connection may be spread over several such connections. The data is then said to be split and recombined.

Where a requested quality of service cannot be met at all, then the transport connection will not be established. The session layer then has the choice of requesting a lower quality of service or of notifying the presentation layer. The transport layer will notify the session layer during the session if the agreed quality of service cannot be sustained.

## 6.2.5 Network layer

This layer is responsible for the routing and switching of data between transport entities, and can be seen as providing the network between the open systems. This network may in fact by made up of several sub-networks, possibly of different types with different capabilities. All relay functions are provided within or below this layer.

The features that may be provided by the network layer are familiar, following the earlier chapters on X.25, and include the following:

- *Routing and relaying*
  To route the data across the most appropriate paths between the transport entities.
- *Network connections*
  Using several sub-networks in a chain.
- *Multiplexing*
  Putting several network connections over a single data link connection.
- *Segmenting and blocking*
  Splitting or combining data units for more efficient transmission or to cater for the particular connections used.
- *Error detection and recovery*
  Either using notification of errors from the data link layer or additionally by checking within the network layer.
- *Sequencing*
  Providing for the delivery of data in sequence. Note that whilst this is inherent in X.25 it is not so with other network types such as connectionless.
- *Flow control*
  To ensure that a receiver of data can indicate a temporary inability to receive more data.
- *Expedited data transfer*
  To allow for high priority data.
- *Reset*
  To restore the connection to a known state.

In X.25 networks the presence or absence of many of these features is fixed, but, in general, the transport layer needs to negotiate the features in order that it can ensure the quality of service. The transport layer may need to implement some or all of these features itself in order to match the requested quality of service with the capabilities of the network. Thus, a class 0 transport layer can run over X.25 layers three and two, whereas if a simple LLC(1) is in use then a class 4 transport layer would be needed with a null network layer.

### 6.2.6 Data link layer

The function of the layer is to detect errors that may occur in the physical circuit. It may also correct these errors, and will notify the network layer of errors that it cannot recover. Implicit in this is the responsibility for the correct delivery of bits, with the data presented to the receiver correct and synchronized.

This layer is responsible for the structure of the data since the physical layer only deals with a stream of bits. The manner of creating this structure depends on the network type, and in the case of X.25 Chapter 1 showed that a frame is used. The various frame types and features allow for error detection and some recovery, as well as separating the data transmission into manageable chunks.

### 6.2.7 Physical layer

The physical layer is responsible for delivering a bitstream from one component to another. This layer therefore takes care of the mechanical and electrical interface between the components.

## 6.3 Interconnection of open systems

It is easy to assume that a system that conforms to the requirements of OSI must necessarily connect to and communicate with any other system that conforms — after all, that is the whole point of the seven-layer model. Despite the more enthusiastic claims of some salesmen, this is not the case.

ISO 7498 lays down a framework within which the more detailed standards are written. It can only be said that the (N)-layers in two open systems perform broadly similar functions to each other, not that they will necessarily communicate with each other.

The seven-layer model is like a computer programming methodology in that it provides a structure. Two programs can be written that conform to the methodology and do the same job, but are written in different computer languages. The sub-sections of the two programs are then not interchangeable and cannot communicate.

At each layer of the model there is a whole host of standards, each with differing requirements of the layers that go above and below. This gives rise to the OSI profile, which is simply a selection of standards, one

**Left diagram:**

```
7        A    B
6    A   B   C   D   E   F
5        A   B   C
4        A    B
3        A   B   C   D
2        A   B   C
1    A   B   C   D   E
```

**Right diagram:**

```
7        A  (B)
6    A   B  (C)  D   E   F
5        A   B  (C)
4        A  (B)
3        A   B   C  (D)
2        A  (B)  C
1    A   B  (C)  D   E
```

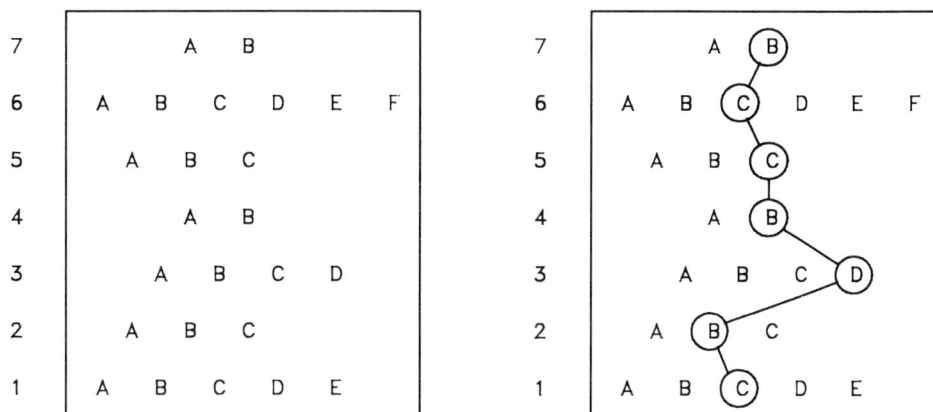

**Fig. 6.4** The seven-layer model with many standards at each layer; a functional cut through the model gives a profile which completely defines an open system

from each layer, that work together to provide an open system. The profile is a complete definition of the open system, and systems that conform to the same profile will necessarily interconnect. The profile can be seen as a functional cut through the layers of the model, as shown in Fig. 6.4. Some practical profiles were discussed in Chapter 3.

## 6.4 Sorting out the X.25 numbering

Chapter 4 introduced the topic of X.25 addresses and the structuring of the address within the available fifteen digits. It is not immediately obvious that a structure is necessary, since there is no problem with each network running its own scheme as shown in Fig. 6.5. The fact that the two networks have an identical address is of no concern, since the addresses are unique within the network.

The problem comes when the two networks are connected together. How does a caller now indicate that access to a service in the other network is required? As long as the caller is a person on a terminal then it is possible for the device linking the networks – the gateway – to ask a question. This is illustrated in Fig. 6.6.

A user on network A makes a normal call to address 916 which is the address of the gateway on network A. The gateway accepts the call in the normal way and then sends data to the screen saying "now where do we connect to?". The user responds with an address, which is the address of the required ultimate destination on network B. The gateway makes a completely separate call to that address, which will hopefully be accepted, and the job of the gateway is then to pass data messages between the two calls.

An alternative to this scheme is for each network to have alias addresses for the services in the other network, and for the gateway to resolve the aliases when a call is made. Thus a user on network B, for example, might call address 8476 which is an alias address and the call is routed to the gateway. The gateway then translates 8476 to the actual

**157**

service address on network A – address 123 perhaps – and makes the call into the destination network. This scheme is expensive since it requires in each network knowledge of all possible services in all possible networks.

Where a network is anything other than extremely parochial, it is better to have a standardized numbering scheme to resolve the addressing confusion. The standard agreed internationally for public data networks is recommendation X.121. This recommendation works in much the same way as telephone numbers, that is it is hierarchical and has fields for country, area, etc., down to the individual subscriber. Unlike the telephone system however, all digits of the X.25 address must always be specified. Making a call within a particular area of a particular country requires that both the country and the area code must be specified. This means that a particular subscriber is always accessed by the same address from anywhere in the world.

Of course, the particular PAD used may allow a shorter form of address to be specified by the user, or indeed a completely different form of address as discussed in Chapter 2. The PAD will, however, form a complete address conforming to X.121 in the call packet that it sends into the network.

X.121 defines a structure for the first 14 digits of the X.25 address, and leaves the final digit unused. The first field of the address is a four-digit code specifying the country, and the particular network within the country. This is called the *Data Network Identification Code* (DNIC). The first digit of the DNIC represents what the recommendation calls a "world zone", but is essentially a continent identifier. The middle two digits represent the country within the continent. And the final digit of the DNIC represents one of the available networks within that country. If a country should have more than ten national networks then further country codes are allocated.

Some examples of country codes are as follows:

| | | | |
|---|---|---|---|
| 206 | Belgium | 334 | Mexico |
| 208 | France | 440 | Japan |
| 234 | United Kingdom | 505 | Australia |
| 310 | USA | 665 | South Africa |
| 311 | USA | 716 | Peru |

Following the DNIC is the identification of the individual subscriber within the network. X.121 imposes no structure on this number except to say that it is a maximum of ten digits. Thus, X.25 addresses in different countries can be different lengths.

In the United Kingdom PSS network, the ten digits are further sub-divided as follows, giving a highly structured and organized numbering scheme. This is an example of how PTTs can extend the spirit of X.121 across the complete address.

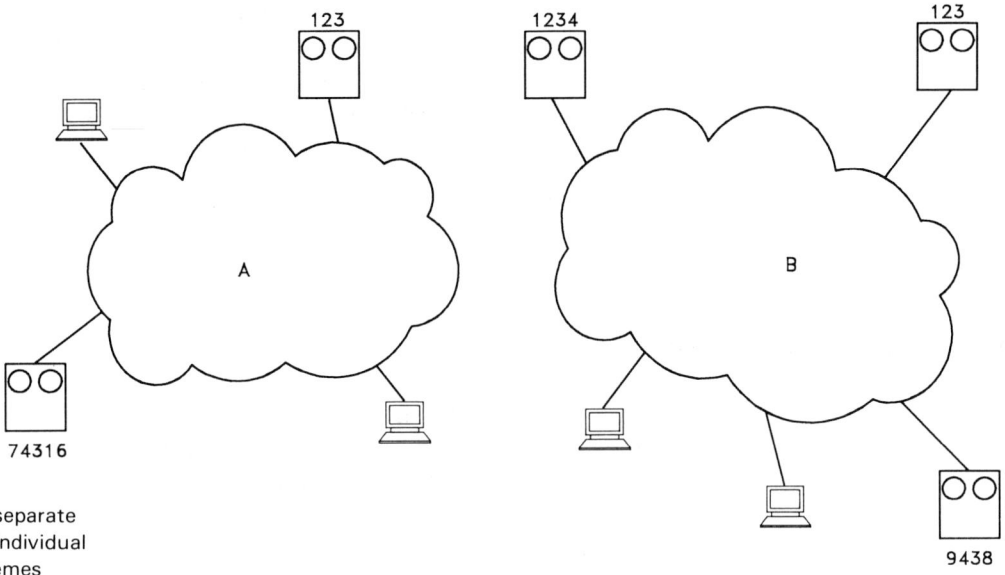

**Fig. 6.5** Two separate networks with individual addressing schemes

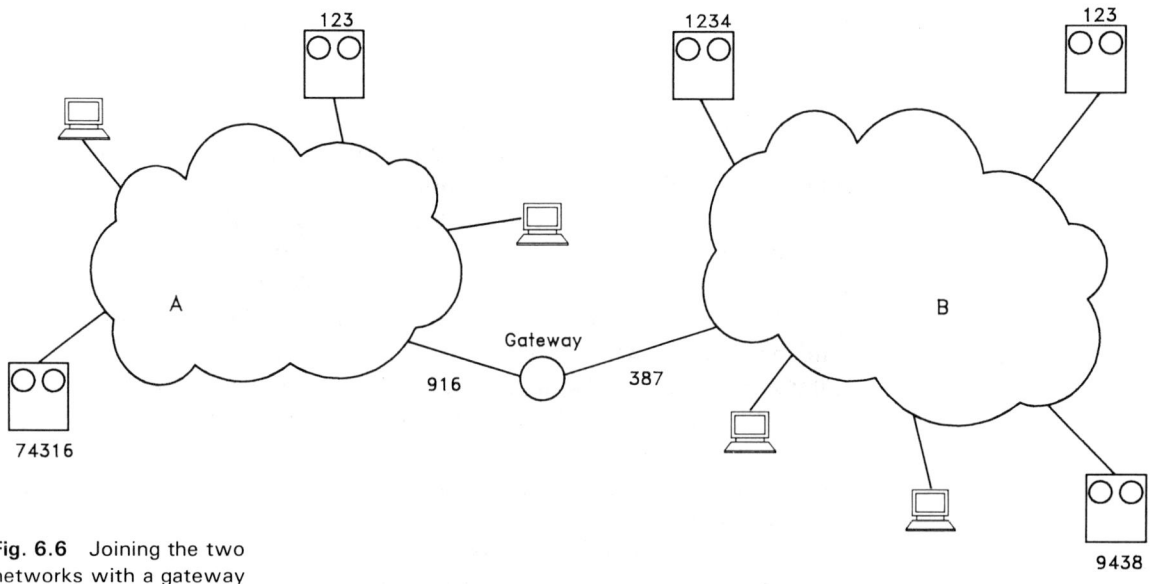

**Fig. 6.6** Joining the two networks with a gateway

First three digits   Area code, based on the same numbering system as is used in the telephone network

Next five digits   Subscriber number

Last two digits   For private use

The final two digits of the PSS number are not used by the network at all. Whether zero, one, or two digits are inserted by the caller, they will be passed unchanged to the destination. The digits can therefore be used for any entirely private purpose between the caller and the service, much as Call User Data is used. The two digits typically specify further routing within the subscriber site. Thus, a subscriber may have a number of services all connected to a single PSS link via a switch, and callers can use the digits as a sub-address to specify which particular service they want.

## 6.5 Connecting networks together

Chapters 1 and 2 have already shown that two X.25 "pipes" can be connected by an X.25 switch, and that a sufficiently configurable switch can manipulate the addresses passed through it. Given two X.25 networks therefore, a switch seems to be the obvious component to link them together. This is especially true when X.121 is considered and the problems of address translation are eased.

It will be recalled from Chapter 4 that X.25 is a recommendation that specifies the interface between a DCE on a public network and a DTE using that network. It does not specify how data is represented within the network. So, whilst X.25 is the defined way of accessing the network, there is no reason to suppose that it is the best way of joining two networks.

Connecting two network DCEs together by a switch is feasible, and for many applications perfectly adeqate. However, for large-scale applications, such as linking two national networks together, a protocol designed for the application will prove more effective. One such protocol is defined in recommendation X.75, and this interfaces to each network at a *Signalling Terminal Exchange* (STE). The use of the STE and X.75 protocols ensures a much greater traffic throughput than would be possible with a single switch, for reasons explained below.

Figure 6.7 shows how a number of X.25 networks might be connected. Within each X.25 network a user and a service each communicate with the network using purely X.25 procedures. They are unaware of STEs and X.75 in the same way that they are unaware of other DTEs on the network.

If a user on one network accesses a service on another network, then both communicate with their respective networks using purely X.25 procedures, and are unaware of any further complexity. In this case though, each network has to communicate directly with the other using X.75 protocols in order to pass the data between them. A network may also simply be an X.75 routing system, passing data between other networks but with no DCEs of its own being involved.

It is emphasized that in a call involving X.75, the DTEs follow exactly the same procedures as they would for a call within their own network, except that the DNIC of the address will specify another network. All

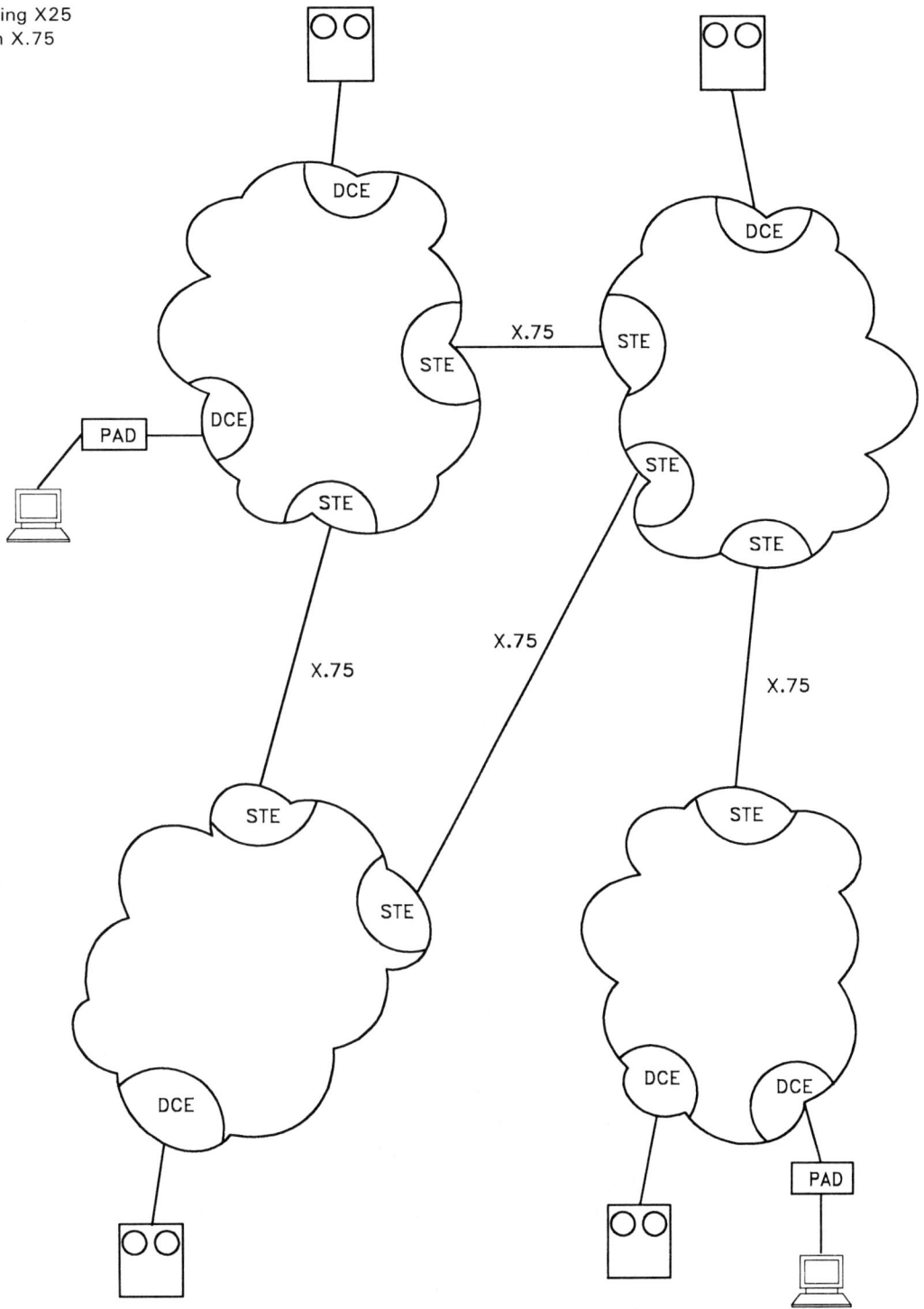

**Fig. 6.7** Joining X25 networks with X.75 protocols

the work of routing the data to the STE, and of using X.75 protocols, is the responsibility of the network.

The X.75 protocol is similar to that of X.25, but is designed with the aim in mind of bulk data transfer. The following factors are especially relevant to this:

- The physical links are expected to run at the high speed of 64 K bps, though other speeds are allowed.
- The multilink procedure (MLP) described in Chapter 4 is used, so that the traffic between the two STEs can be spread over multiple physical circuits.
- Modulo-128 frame counts are used to compensate for the long propagation delay that would be expected in international circuits, by allowing larger numbers of frames to be inflight at the same time.
- The Call procedure (call request, call connected, clear request) is extended from that of X.25 to include network utilities requests. These requests are a means for STEs to communicate, and some of them are listed below.

*Transit network identification*  To indicate networks on the route between the calling and called networks.
*Call identifier*  To give a unique name to the call for identification.
*Throughput class indication*  To specify throughput classes.
*Window size and packet size indication*  These apply to the interface between the networks.
*Transit delay indication*  Allowing the overall propagation delay of the call to be assessed.

## 6.6  The IBM Systems Network Architecture

Systems Network Architecture (SNA) is the approach adopted by IBM for linking their services and users together. So, using SNA, any terminal on a network can access any service on the network, and software in any two hosts can communicate. SNA has undergone a great deal of development since its introduction in around 1974, and the networking complexity has grown from a strictly hierarchical host-to-terminal tree into a system that allows full interworking.

SNA is a layered protocol system and, whilst it does not exactly align with the OSI recommendations, it does offer a similar level of service to that of an open system.

### 6.6.1  The basic architecture

Terminals attach to an IBM host via a Front End Processor (FEP). The

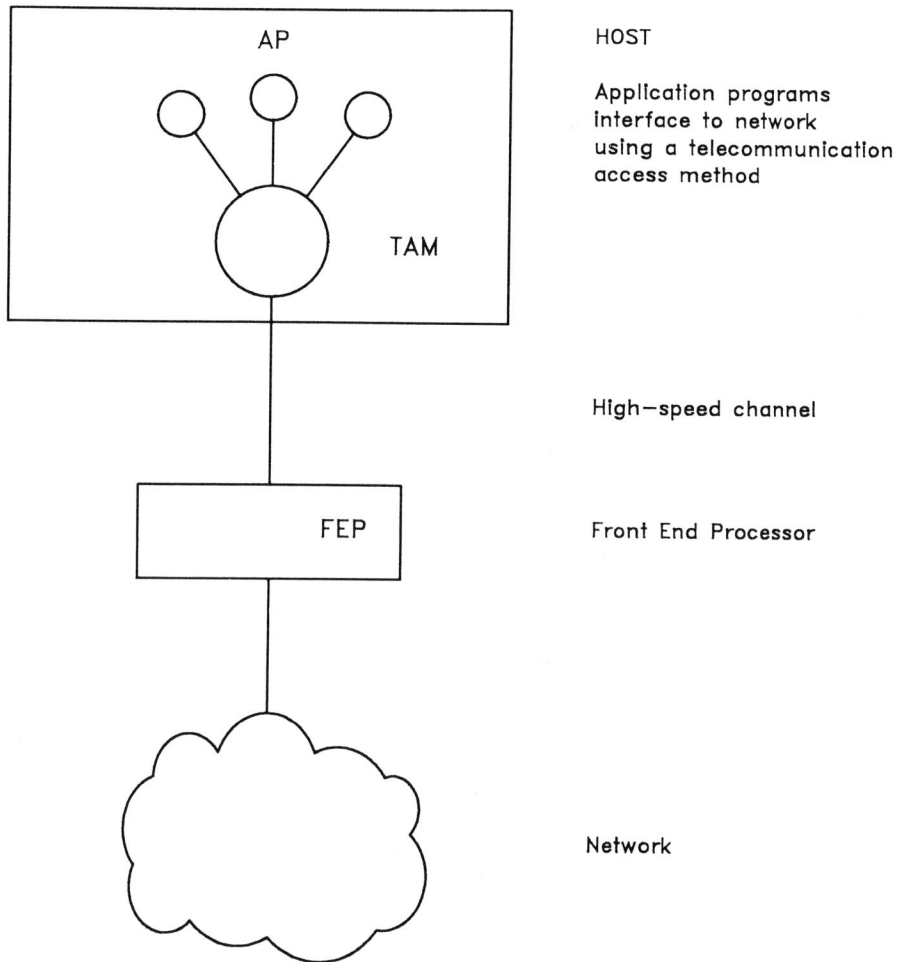

**Fig. 6.8** Basic architecture of a SNA network

FEP is responsible for all of the communications aspects of the system, and connects to the host via a high-speed channel. The host software in control of this channel provides a telecommunications access method for the applications running within the host. This is shown in Fig. 6.8.

Within the host the software may be Advanced Communications Facility / Virtual Telecommunications Access Method (ACF/VTAM) or Advanced Communications Facility / Telecommunications Access Method (ACF/TCAM).

The Front End Processor is also referred to as a Communications Controller, and runs the Advanced Communications Facility / Network Control Program (ACF/NCP). Typical models of FEP are 3705, 3725, and 3720.

A special option of ACF/NCP is the Network Packet Switched Interface (NPSI), which allows links from the FEP to run X.25 protocol. The normal physical links to the FEP run Synchronous Data Link Control (SDLC) protocols.

Communication in an SNA network is between two Network Addressable Units (NAUs). An NAU has a unique address on the network and can both initiate and receive a call. An NAU may be assigned to four classes of device:

- Host computers, more specifically to the Systems Services Control Point (SSCP) which is part of the ACF/VTAM or ACF/TCAM software.
- Front End Processors.
- Cluster controllers such as the 3274, which control a number of terminal devices.
- Terminals.

This gives rise to the type of network shown in Fig. 6.9.

This topology has many of the elements of the old strict hierarchy visible, but nevertheless does provide a great deal of flexibility. Each host, or more strictly each SSCP, "owns" the items beneath it, which are referred to as its domain. Thus each terminal, cluster controller, and front end processor are in a single SSCP domain.

Communication with a service in a different domain is possible, and is performed by asking the owner SSCP to set up the path. The owner SSCP does this by asking the owner of the called service for a path and linking across the two front end processors. Each owner SSCP continues to control the entities beneath it for the session, but the session data does not go through the owner host unless it needs to.

## 6.6.2  The SNA node

A node can be any one of the components outlined earlier, whose functions clearly depend on which type of node it is and how it is connected to the network. Within the node there will typically be a single Physical Unit (PU), and a number of Logical Units (LUs). Each PU and LU are Network Addressable Units so have an identity of their own on the network.

The Physical Unit is a piece of software which handles the SNA trunk link(s) for the node and is generally concerned with global functions of the unit.

The Logical Unit is a piece of software which handles an individual network user such as an application program or an interactive terminal. All communication between users of the network involves two LUs, and the LU undertakes all interfacing between the person or program and the network. The Logical Unit provides a standard end-point for the network to deal with, and the network does not have to deal with the subtleties of the actual end-user.

Communication between Logical Units is via a layered hierarchy that has the same philosophy as the seven-layer model. Within the Logical Unit are three layers:

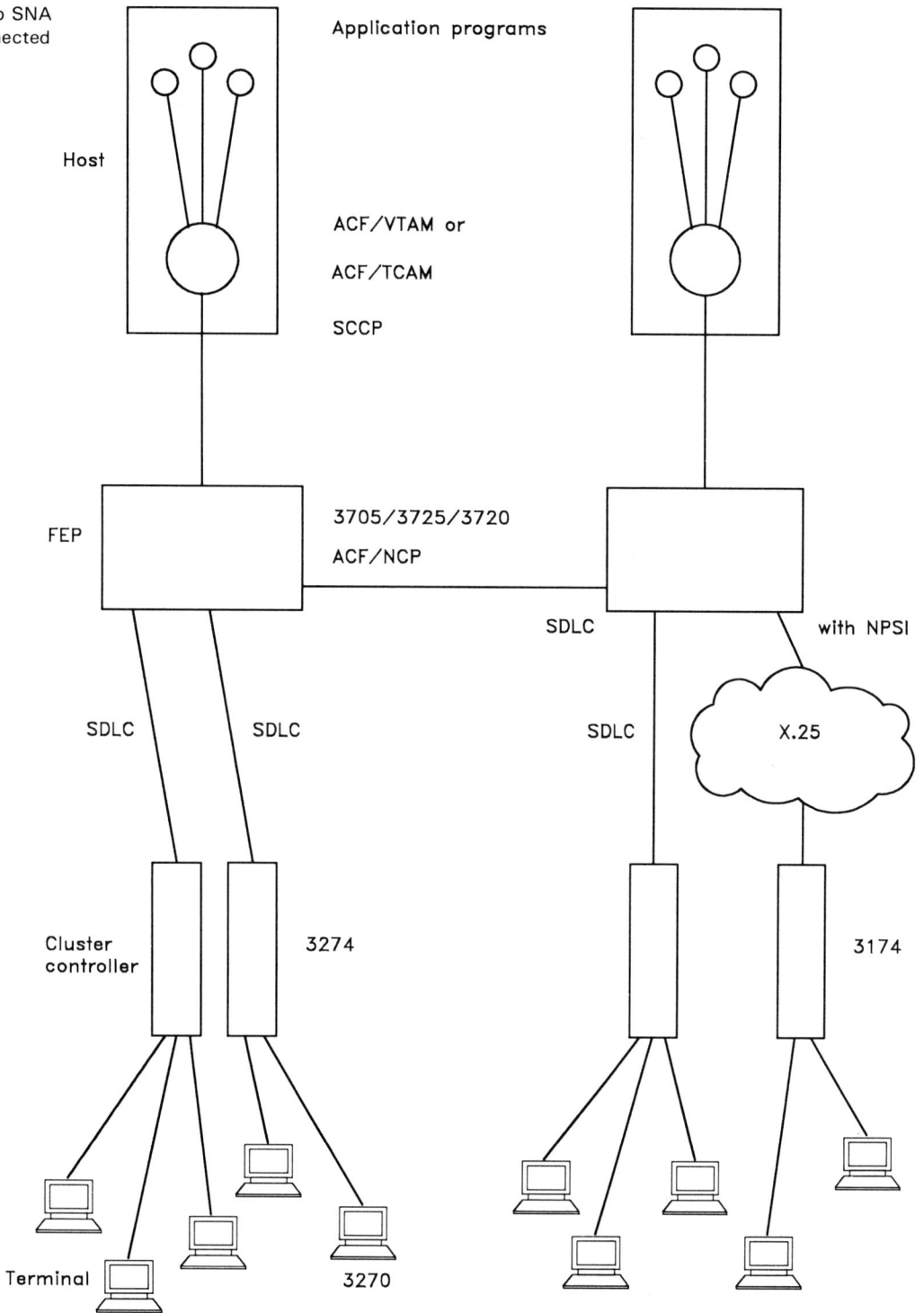

**Fig. 6.9** Two SNA domains connected together

Application programs

Host

ACF/VTAM or

ACF/TCAM

SCCP

FEP

3705/3725/3720

ACF/NCP

SDLC

with NPSI

SDLC

SDLC

SDLC

X.25

Cluster controller

3274

3174

Terminal

3270

- *Function Management Data Services*
  Here the precise service required by the network user is made available, and the particular style and capabilities of the user are standardized so that it can be interfaced to the network.
- *Data Flow Control*
  This layer manages the session between the two users, providing message sequencing, matching of responses to requests, and control of when the users may transmit data.
- *Transmission Control*
  This is responsible for multiplexing traffic from a number of sessions over a particular network path.

The data interface below the Transmission Control layer is in the form of Basic Information Units (BIUs). These contain protocol information from the upper layers – and possibly data from the users as well – and contain all the information necessary to route the BIU through the network. This routing, known as Path Control, is performed in two layers:

- *Path Control*
  Here the actual path through the network is selected. Multiple links can be used to increase the effective throughput, and the layer takes care of message sequencing.
- *Data Link Control*
  This provides error-free data transfer across a physical communications link. Over much of the network this will be by SDLC which is broadly similar to X.25 layer two.

**Fig. 7.1** The elementary electrical circuit

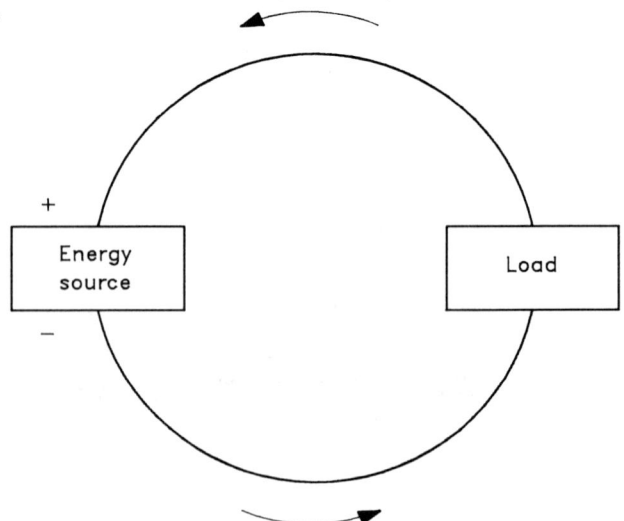

# 7 Plugs and wires

## 7.1 Introduction

This chapter aims to resolve some of the confusion that surrounds the physical side of connecting things together. It ought to be very simple since most devices have standardized on the use of a twenty-five-way connector, but a glance in the catalogue of a cable manufacturer shows that there are a number of ways of making even the simplest of connections.

Then there is the problem of the recommendations. Some manufacturers claim V.24 conformance whereas others claim RS-232. Still others talk about X.21 and X.21bis, and even X.20. The chapter will explain what all these standards actually say, and how they link together.

The chapter also looks at the circuit inside of the network components to illustrate what all the little switches and jumpers actually do, and why they are there.

## 7.2 Physics

To understand what actually happens on the piece of wire between two components, it is instructive to go back to the electrical basics. The elementary circuit is shown in Fig. 7.1. The circuit consists of three things:

- A source of electrical energy, such as a battery or the power supply of a piece of equipment. The power source has two connections on it marked ( + ) and ( − ) which have a differing amount of electrical energy on them. The electrical energy is called potential; thus there is a potential difference between the two connection points. This potential difference is measured in volts.
- A load, which is the thing that is to be powered by the energy source. This may be anything, such as a light bulb, a dishwasher, or a computer.
- Conductors, which can conduct electrical energy from any point to any other, like pipes conduct water in a plumbing system.

The key point to note is that the two connections of the power source are joined together by the conductors and the load. There is a continuous route, or circuit, from the ( − ) connection, through the load, to the ( + ) connection.

Now, because the two power source connections are joined, and because they have a different electrical potential, electricity will flow from one connection through the load to the other connection, to try and equalize the potential. It is like water flowing along a pipe in order to try and equalize the pressure at either end. The flow of electricity through the load will activate it – the bulb will light, the dishwasher will wash, or the computer will run.

**Fig. 7.2** Electrical circuit for data communications

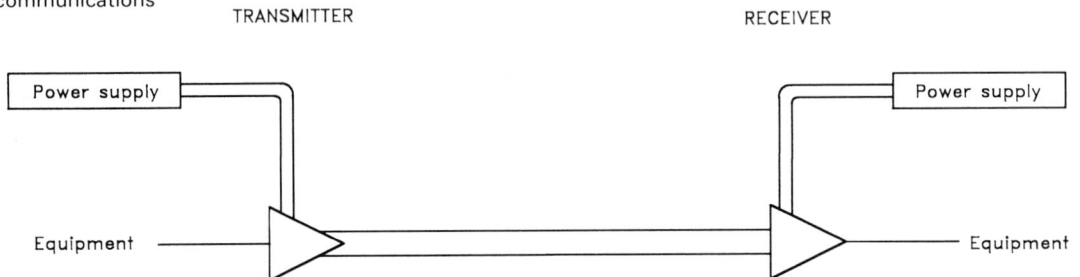

In the case of an electrical circuit between two components of a data network, the diagram can be drawn as shown in Fig. 7.2. The transmitter contains a device which is connected to the power supply, and which has two output connections. The device can either keep the outputs at the same potential, or make them a different potential, according to instructions from the rest of the equipment.

Binary data can now be sent to the receiver: no potential means zero; a potential difference means one. In the receiver there is a device which is again connected to a power supply, and which also forms a load across the two incoming conductors. If electricity flows through the load then the rest of the equipment is informed. The receiver can thus detect the binary data sent from the transmitter.

The major flaw in this system is that there is no way of detecting whether a zero is being sent, or if there is a fault such as a broken conductor. Actual circuits, such as those defined in V.28, address this problem as discussed later in this chapter.

## 7.3 Transmission types

The previous section showed how a bit can be passed from the transmitter to the receiver, but did not lay down any transmission procedures. The need for procedures is clear if consideration is given to sending two consecutive one bits. How does the receiver know that it is two bits rather than a single bit on for a long time? There has to be either an element of timing involved, or some means of indicating to the receiver when it should take note of the transmission. These two procedural transmission methods are called Asynchronous and Synchronous respectively.

### 7.3.1 Asynchronous transmission

In asynchronous transmission both the transmitter and receiver have internal clocks which are set to run at the same speed. Every time the clock ticks then the transmitter sends a bit, either by leaving the line in the state that it is, or by changing the state if this bit is different from the last.

For example, suppose it is required to transmit 01101001, the procedure would be:

| | |
|---|---|
| 1st tick | Ensure line is transmitting zero |
| 2nd tick | Change line state to transmit one |
| 3rd tick | Leave line transmitting one |
| 4th tick | Change line state to transmit zero |
| 5th tick | Change line state to transmit one |
| 6th tick | Change line state to transmit zero |
| 7th tick | Leave line transmitting zero |
| 8th tick | Change line state to transmit one |

At the receiving end, every time the clock ticks the receiver determines the state of the line and informs the rest of the equipment.

There are two major problems with this procedure:

● Even if the two clocks could be started in synchronization, they would inevitably drift out of synchronization and cause the receiver to look at the line when the transmitter had not ensured that it was correct.

● What happens when there is no data to send?

These problems are resolved in asynchronous procedures by framing the transmission. The bits to be sent are split into "chunks" of typically eight bits each, and these are sent each framed by a start bit and a stop bit. The transmission may appear as shown in Fig. 7.3. The gaps between the frames are of arbitrary length and depend only on when the transmitter has data that it is required to send. In the gaps the line is always in the one state.

**Fig. 7.3** Asynchronous framing by Start bit and Stop bit

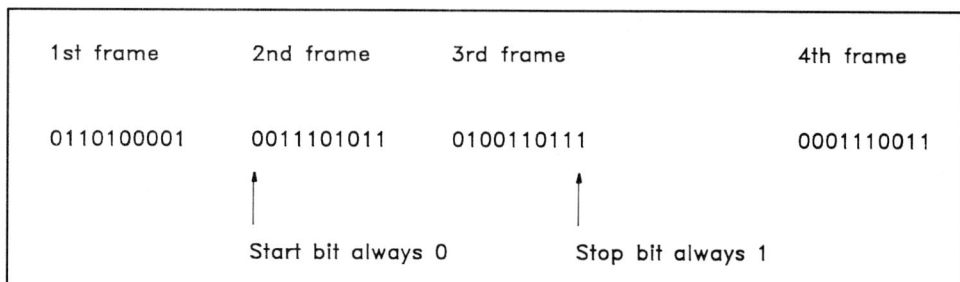

| 1st frame | 2nd frame | 3rd frame | 4th frame |
|---|---|---|---|
| 0110100001 | 0011101011 | 0100110111 | 0001110011 |

Start bit always 0      Stop bit always 1

When the transmitter has accumulated eight bits to send, then it sends a zero on a tick of the clock, followed by the eight bits on subsequent ticks, followed by a one on the tenth bit. The transmitter then remains idle until the next eight are ready to be sent. The leading zero is the *start bit*; the trailing one is the *stop bit*.

The receiver monitors the state of the line during the gap, and as soon as a zero is detected it will start its internal clock. At the next eight ticks the state of the line is detected and the data bits received. At the final tick the receiver can check that the line has gone back to the one state.

The stop bit is not really necessary as a check since it is unlikely that the receiver clock will go out of synchronization in ten bit times; however, it may be useful in detecting errors such as mismatched clock speeds or in detecting break signals.

*A break signal* is the transmission of the zero state from more than ten bit times, and therefore contravenes the normal framing procedures. It follows that it is easy to detect in the hardware of the receiver. Break is used as an attention-grabbing mechanism, for instance between a terminal and a PAD. It is useful because it is not a "proper" character, and therefore does not affect the transparency of data transfer.

The length of the break signal is dependent on the time it takes to transmit ten bits, and therefore on the speed of transmission. Clearly the break will still be detected if the signal is longer than the ten bit times, and some devices use a fixed time for the signal whatever the line speed. This fixed time is commonly 135 ms which is satisfactory for line speeds of 75 bps and above. (Note that at 75 bps just over 10.1 bits can be sent in 135 ms.)

The stop bit is often useful to mechanical printing devices in that it gives them time to prepare for the next character, which may immediately follow the stop bit. In fact, with low-speed devices it is common to use one and a half or two stop bits to give extra time for preparation. One and a half stop bits sounds an extremely "un-digital" number, but it should be remembered that the stop bit is simply a guaranteed idle time before the next character, rather than an information transfer mechanism.

Although there is no reason why framing lengths other than ten (start, stop, and eight data) bits should not be used, this is the standard since most data communications devices generate eight-bit bytes. The coding of the eight-bit bytes is normally in ASCII or International Alphabet Five (IA5), which are shown in Appendix C.

Asynchronous framing is cheap to implement since it only requires a single transmission circuit between transmitter and receiver; however, it is inefficient since only eighty percent of the bits are true data. The procedure is normally used with terminals and printers where data is relatively sparse and the inefficiencies are not significant.

### 7.3.2 Synchronous transmission

In synchronous transmission there are two circuits between the transmitter and receiver: the data line and the clock line.

The *data line* is used for the zeros and ones of the data as shown in the previous section for asynchronous transmission. However, the data has no structure in synchronous transmission − it is a continual stream of bits. This means that if there is no real data ready for transmission, then some form of padding data must be sent to keep the line busy. There can be no gaps or idle states in this form of transmission.

The *clock line* carries the ticks of a clock from one component to the other, so that the transmitter can set the state of the line, and the receiver can detect the state, using the same clock. This guarantees keeping the two ends in synchronization.

The procedure for sending 0110 is shown in Fig. 7.4.

**Fig. 7.4** Procedure for synchronous transmission of 0110

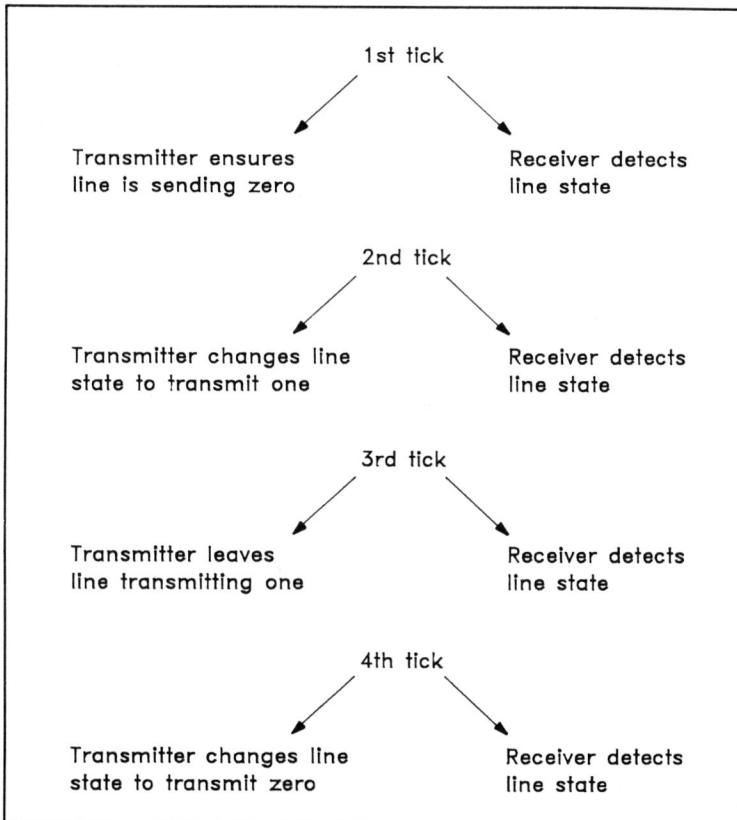

The receiver will normally detect the state of the line in the middle of the bit time to ensure that the transmitter has had a chance to set the state. It does not matter which of the transmitter or receiver contain the clock. It can be in the receiver, in which case the clock line transmits in a different direction to the data.

Synchronous transmission is expensive because it involves double the hardware of asynchronous – two transmitters, two circuits, and two receivers. The advantage is that everything that is transmitted is real data, and therefore the procedure is efficient.

Synchronous transmission is typically used on trunk routes in data networks – X.25 lines for example – where speed and therefore efficiency are fundamental requirements. In the case of X.25, if there is no layer two data to send then flag bytes will be transmitted as shown in Chapter 1. Different padding techniques will be used for different network types.

### 7.3.3 Manchester encoding

This is a synchronous method of transmission that only requires a single transmission path. In this method, the clock and the data are carried together on the transmission path and have to be separated by the receiver.

Figure 7.5 shows an example of activity on the circuit in the form of a graph showing signal against time. The signal is made up of pairs of bit values, that is each pair of bits of the signal represent one data bit. The value of the data bit, zero or one, is represented by the transition of the signal between the two signal bits.

- A transition from one to zero in the signal represents a zero data bit.
- A transition from zero to one in the signal represents a one data bit.

**Fig. 7.5** Manchester encoding showing two signal bits for each data bit

Thus we can see that Fig. 7.5 represents 00100.

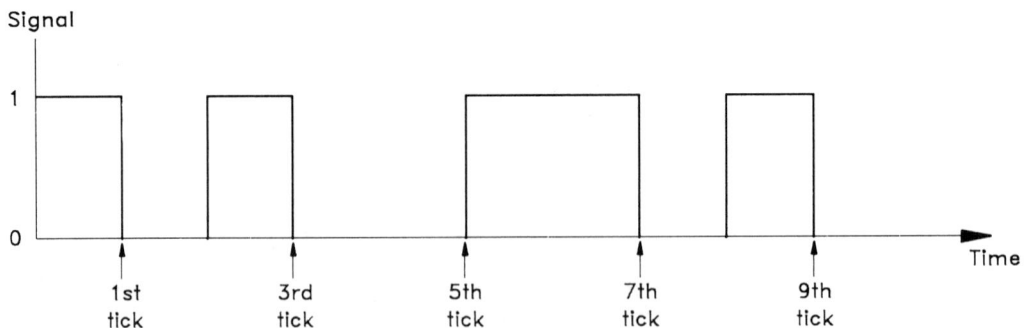

Because each data bit is represented by two signal bits, the transmitter can always create a transition in the appropriate direction whatever the current state of the circuit. Where the transition occurs, then, as well as determining the data bit, the transition marks the position of the clock tick.

Manchester encoding is clearly a fairly complex method of transmission compared to separate clock and data systems, but the use of a single circuit is especially useful in some circumstances. These are mainly concerned with local area networks that use co-axial cable, where there can only be a single circuit.

## 7.4 Serial and parallel transmission

Whether asynchronous or synchronous transmission is used, the rate of transmission can be improved by using a number of signal paths in parallel. Clearly, if two transmission paths are used, then double the number of data bits per second can be sent.

Normally, when a number of paths are used in parallel, then the number allows a whole character or computer word to be transmitted in a single tick of the clock. In data communications where there is some standardization on eight-bit characters, then parallel data paths usually have eight circuits in them.

When a single circuit is used then each bit of the transmission must go singly. This is called serial transmission.

## 7.5 Simplex and duplex

A one-way data path between two components is called a *simplex* connection. This would have a receiver at one end of the circuit and a transmitter at the other, and might be used between a computer and a printer. Data transmission in the opposite direction is not possible.

A connection between two components that has two paths, so allows data to be transferred simultaneously in both directions, is called a *full-duplex* connection. This obviously involves double the cost of a simplex connection because there is twice as much transmission and reception equipment as well as the two data paths. This is nevertheless the most common type of connection.

It is possible to have a half-duplex connection which is a single path between two components, but allows data to be transferred in each direction, though only one way at once. This type of connection only saves on the cabling costs, since the equipment at either end must be more complex than for a full-duplex connection. There must also be a protocol between the two ends to decide who can send when.

Some network protocols are inherently suitable for use with half-duplex transmission. They generally operate a poll-select mechanism where a master station polls around all its subordinates and asks whether they have anything to send. If they have, then they are allocated the transmission medium.

## 7.6    Signalling rate and data rate

The *signalling rate* is the number of transitions per second that can be sent on the data path(s) between transmitter and receiver. It is normally expressed in bits per second (bps).

The *data rate* is the amount of data that can be transmitted per second between the transmitter and receiver. This is known as the Baud rate and may also be measured in bits per second.

On a synchronous circuit with one data path then only one data bit can be transmitted on each clock tick, and the data rate is thus the same as the signalling rate.

On a synchronous circuit with eight parallel data paths, then if the signalling rate is 1200 bps, the Baud rate is 9600 bps. This is because in each bit transition, eight data bits are transmitted.

On a circuit using Manchester encoding the Baud rate is half of the signalling rate because it takes two signal bits to transmit a single data bit.

## 7.7    The two ends of the circuit

The two pieces of equipment connected together by one or more circuits are not equal. One provides a networking service, and one is a user of that service. This concept originated in the use of modems as shown in Fig. 7.6.

As far as the application is concerned – the terminal using the computer – the modems and any intervening network form a simple pipe along which data can pass. The connection between the terminal and the modem is between a user of the pipe, or network, and the provider of that service. The connection between the computer and the modem is likewise between a user of the network and the provider of the service. Although the computer supplies a service to the terminal, it can only do so by using the network.

The provider of the network service is called the *Data Circuit Terminating Equipment* (DCE). The user of the networking service is called the *Data Terminal Equipment* (DTE). To be more accurate, these are the names given to the interface presented by the equipment, rather than to the whole component. Fig. 7.6 can now be redrawn as shown in Fig. 7.7.

**Fig. 7.6** Modems providing networking service to both user and computer

Computer                                                     User of computer

Modem                          Modem

Fig. 7.7 Location of DTE and DCE interfaces

- It is a fundamental principle that a DTE always connects to a DCE.
- Items of data communications equipment always present either DTE or DCE interface(s). No other type is defined.
- The DTE-DCE connection is effected by a cable that is purely passive.

Whilst modems are always DCEs because their inherent function is to provide a networking service, other items of equipment can equally be provided with either DTE or DCE interface(s). Thus Fig. 7.8 shows a computer with DCE interfaces to which terminals with DTE interfaces can connect directly.

Fig. 7.8 Terminals connected to computer by DTE-to-DCE interface

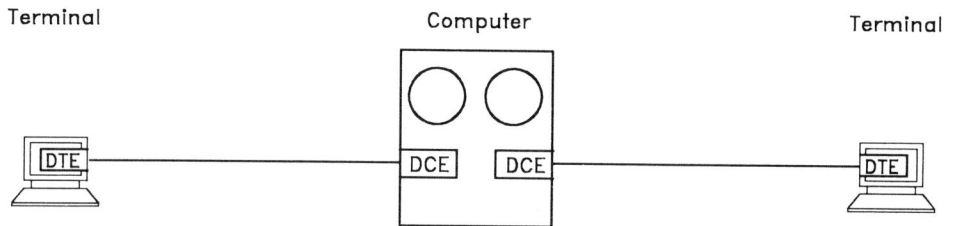

If the length of the cable between the terminal and computer becomes too great for it to form an effective circuit, then the situation could be remedied by using modems. However, there is a difficulty because the DCE of the modem could not connect to the DCE of the computer. It would be necessary to provide a DTE interface on the computer as shown in Fig. 7.9.

Fig. 7.9 Computer needs DTE interface to connect to modem DCE interface

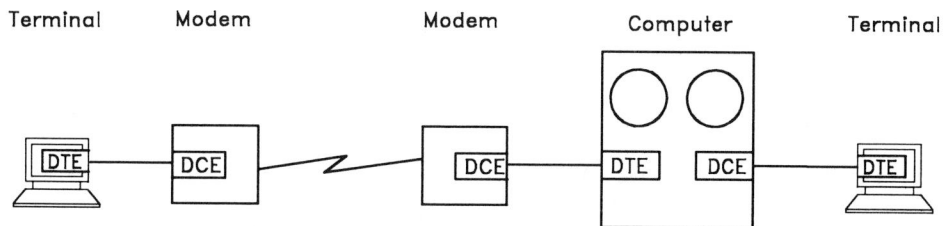

The connection between the two modems is not a DCE-to-DTE one because the modems are necessarily equal partners in providing the connection. This is instead defined in one of a series of recommendations, and the one used depends on the desired speed of transfer and the available physical circuit. Some examples are shown below:

175

- V.21    300 bps over public telephone network
- V.22    1200 bps over two-wire circuit
- V.26    2400 bps over four-wire circuit
- V.32    9600 bps over two-wire circuit
- V.33    14 400 bps over four-wire circuit

The circuit types referred to in the recommendations are the standard offerings of PTT and third party suppliers.

The terms DTE and DCE are also used in X.25 layer two to identify the user and supplier of the X.25 network. The two sets of DTE and DCE associations are completely independent, though they may apply identically to a particular piece of equipment as shown in Fig. 7.10. In this situation the user is an X.25 DTE, that is sends SABM frames, and also has a DTE interface at the physical level to achieve the physical connection to the network. Note that it is usual for the X.25 network administration to supply both modems.

**Fig. 7.10** Separation of layer one and layer two DTE and DCE

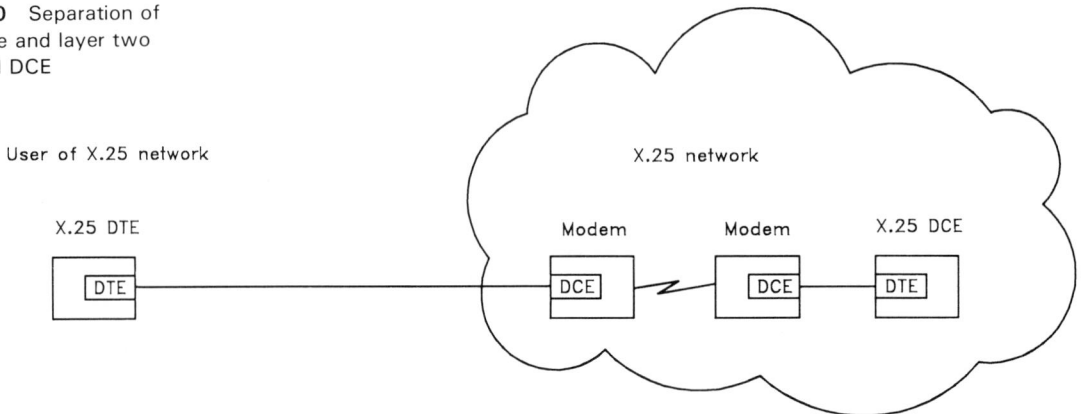

If the user was actually a PAD, then it would in turn provide connections for terminals to actually use the X.25 capabilities. These connections may be either DTE or DCE as shown in Fig. 7.11. In a PAD, reference is sometimes made to DTE-P and DCE-P to indicate explicitly the packet, or X.25, side, and DTE-C and DCE-C to indicate the character or asynchronous side.

The DTE-to-DCE connection has three important aspects which must be defined if the connection, the circuit, and data transfer are to succeed:

- What circuits are going to exist across the interface? This covers whether data is passed in serial or parallel form, where the clocks come from for synchronous transmission, and how many extra circuits are provided for the DCE and DTE to communicate. The most popular recommendation for this is V.24.
- What connectors are used? The physical details of the plugs and sockets must be defined in order for equipment from different manufacturers to connect together. Also, the allocation of the

circuits to the connector – which pins are used for which signal paths – must be standardized. The most popular recommendation is ISO 2110 which defines the familiar twenty-five-way D-type connector.

● What electrical standards are used? This defines the electrical characteristics of the circuits so that the two pieces of equipment can be connected without damage or danger, and so that reliable data transfer can be achieved. Part of this standardization, or at least a result of it, is the maximum distance allowed between DCE and DTE before data corruption becomes significant. This therefore determines at what point modems become necessary. The most popular recommendation is V.28.

Although it is possible to pick-and-mix recommendations in these three areas, it is usual for a set of theee normally to be used together and rarely used apart. Thus the combination of V.24, ISO 2110 and V.28 is the most common DTE-to-DCE definition in use, and any other definitions use their own three recommendations. This causes some confusion. There is a common belief that V.24 is a complete definition of the interface whereas in fact it is a shorthand reference to the set of definitions.

Many recommendations cross-refer to each other. For example, X.21 defines a set of interchange circuits and the procedures to use them, and refers to X.26 for the electrical characteristics and ISO 4903 for the connector.

**Fig. 7.11** DTE and DCE interfaces throughout network

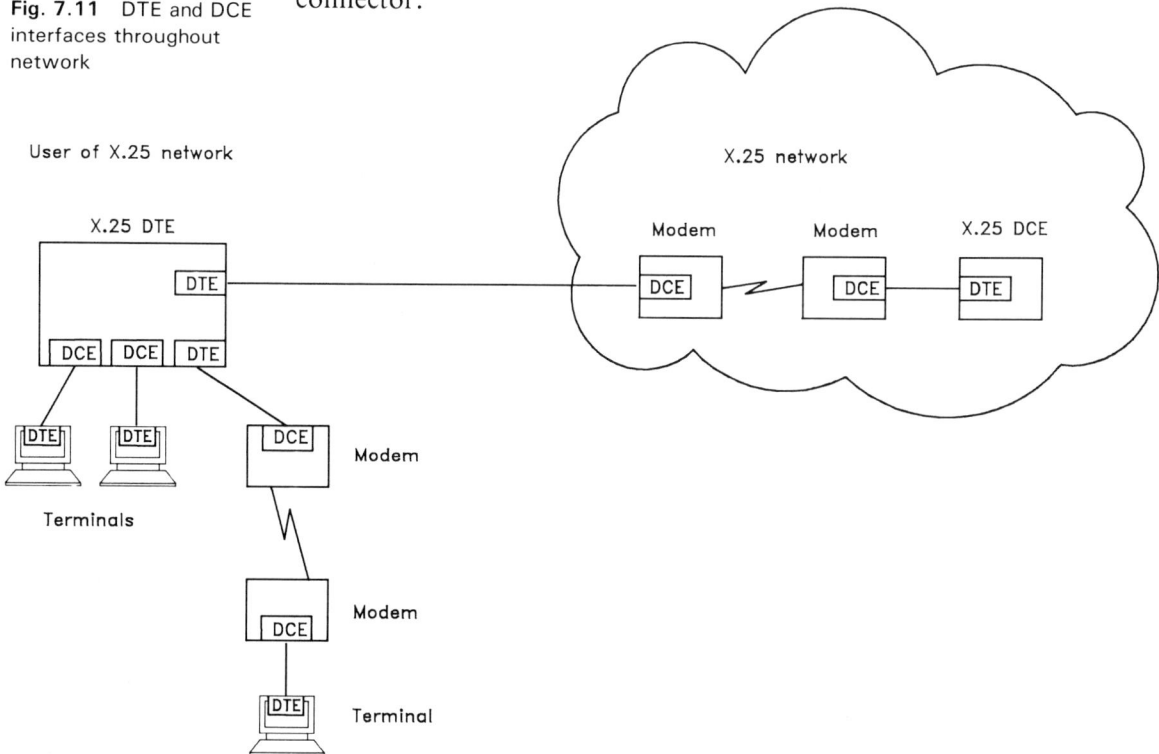

177

This combining of the recommendations is also used in the USA in the RS series of recommendations published by the Electrical Industries Association (EIA). These standards include complete definitions of all aspects of the interface and have the advantage of including a revision letter in the title. This makes them much easier to quote than a recommendation and a date. One of the most popular of these standards is EIA RS-232-C, which is very similar to the V.24/ISO 2110/V.28 combination.

## 7.8   V.24

V.24 is a recommendation governing the signal paths or circuits between a DTE and a DCE. V.24 does not define the mechanical and electrical standards, but does refer to some recommendations for these including ISO 2110 and V.28.

The recommendation defines two sets of interchange circuits which it calls the 100-series and the 200-series. The 100-series consists of forty three circuits which provide the ability to transfer data and a range of control information. The 200-series consists of twelve circuits which allow the DTE to control Automatic Call Equipment connected to the telephone system.

These circuits cover most eventualities required in any data communications environment, and for most general applications provide far more circuits than are ever needed. A subset of V.24 circuits is therefore normally used.

ISO 2110 defines a subset of the circuits, and is the subset in common use in the data communications industry. Even this subset is often reduced further to less than a dozen circuits in some applications. The ISO 2110 subset is listed below. Note that the signal names are always given with respect to the DTE; thus "receive data" is data received by the DTE, and therefore transmitted by the DCE.

The headings in the following list are arranged as follows:

**V.24 circuit number**   *Common name* (common abbreviation)
*Actual V.24 name* if different

**Circuit 102**   *Signal Ground* (GND)
As explained at the beginning of this chapter, two conductors are required to connect the power supply to the load. Because the same power supply is used for each of the circuits between the DTE and DCE, one of the conductors in each circuit can be made common as shown in Fig. 7.12.

The use of this common conductor means that fewer conductors are required and the cable is therefore thinner and cheaper. The conductor is called Signal Ground or Common Return and is circuit 102 of V.24.

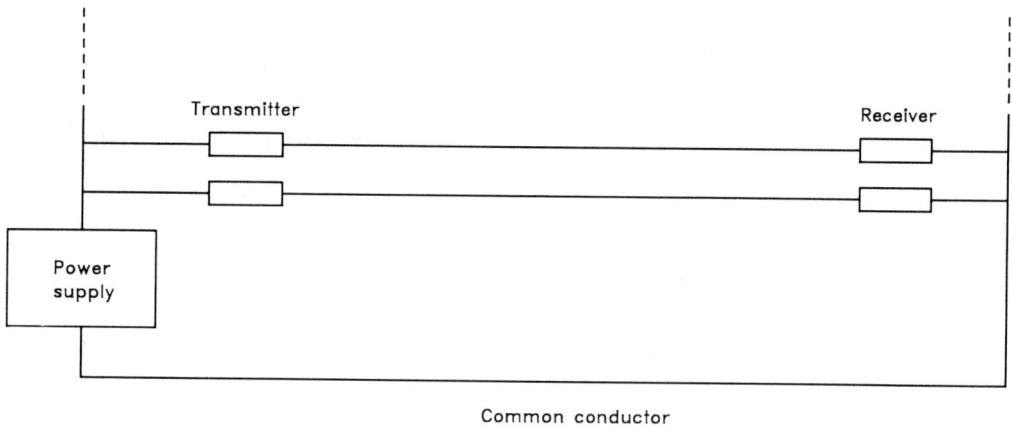

Common conductor

**Fig. 7.12** Connection consists of multiple single conductors for each circuit, and a common conductor for all circuits

This is not a circuit in the same sense as the others, it instead provides the means for the other conductors to actually be circuits.

V.24 allows several types of Signal Ground, but the common subset uses a single conductor which is shared by the transmitters at each end. This is shown in Fig. 7.13. This conductor can be considered as the reference point against which the other conductors are compared. It is the potential difference between the main circuits and circuit 102 that is significant and which determines whether the receiver detects a zero or a one.

**Fig. 7.13** Two-way communications with shared common ground

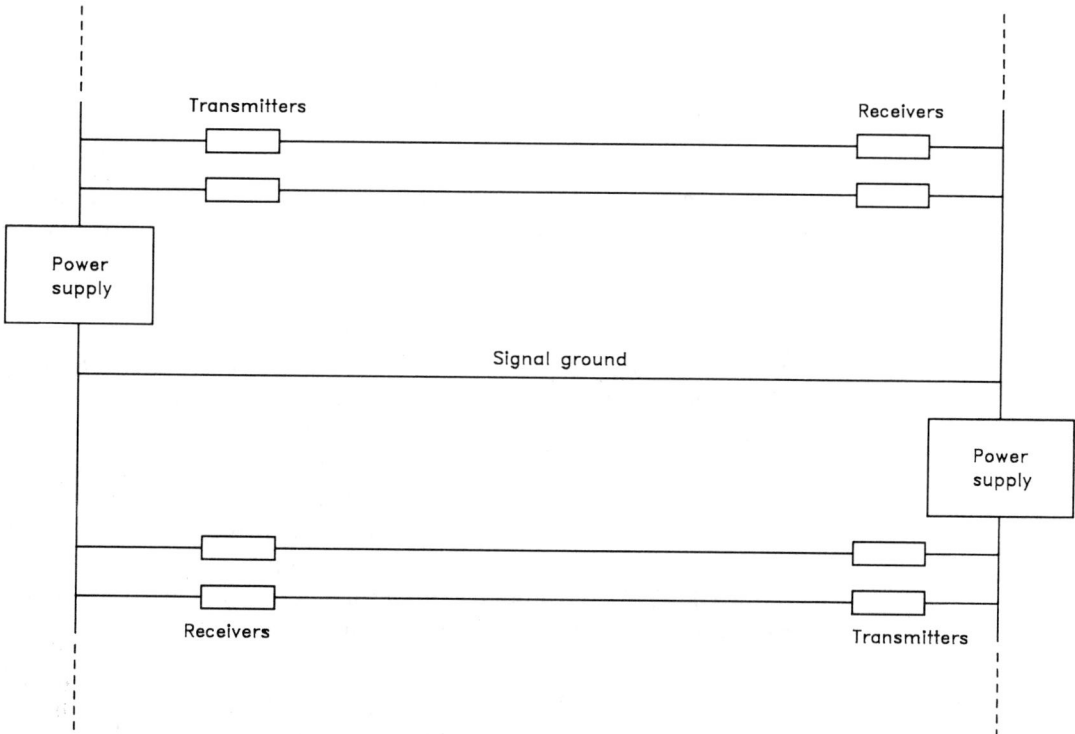

ISO 2110 also provides a conductor called *Protective Ground* (PGND) which is used to connect the chassis in the two items of equipment together. This protects against electrical faults which might otherwise cause a large potential difference between the chassis and damage the equipment. Protective Ground is not defined in V.24.

Protective Ground is normally connected using the screen or shield of the cable between the DTE and DCE. Much of the electrical noise that the cable is subjected to will be picked up by the screen, and is therefore conducted harmlessly to the chassis without affecting the other circuits.

### Circuit 103   *Transmitted Data* (XMT)

This is one of the data streams carried between the two applications. It is named with respect to the DTE so the direction of this data is from the DTE to the DCE.

- The DTE transmits "Transmitted Data"
- The DCE receives "Transmitted Data"

This is a single conductor so the transmission is always serial, that is one bit at a time. The circuit may be used with either asynchronous data, or with synchronous data in which case a clock circuit is used as well (see below).

### Circuit 104   *Received Data* (RCV)

This is the other of the data streams carried between the two applications. It is named with respect to the DTE so the direction of this data is from the DCE to the DTE.

- The DTE receives "Received Data".
- The DCE transmits "Received Data".

This is a single conductor so the transmission is always serial, that is one bit at a time. The circuit may be used with either asynchronous data, or with synchronous data in which case a clock circuit is used as well (see below).

### Circuit 115   *Receiver Signal Element Timing* (RSET)

This is the clock for the Received Data (circuit 104). It is used only when synchronous data is being communicated between the two devices.

The clock is generated by the DCE, so the Received Data and its clock both have the same source and travel in the same direction.

V.24 defines that the clock circuit will oscillate between zero and one with equal periods of each. The transition from zero to one is the clock "tick" and marks the time at which the receiver should check the state of circuit 104. The transmitter will set the circuit state half a bit time before the "tick", and maintain the circuit state for one bit time. The receiver therefore samples the circuit in the middle of the transmitted bit, and should not be confused by the circuit stabilizing after a possible transition from the previous bit. This is illustrated in Fig. 7.14.

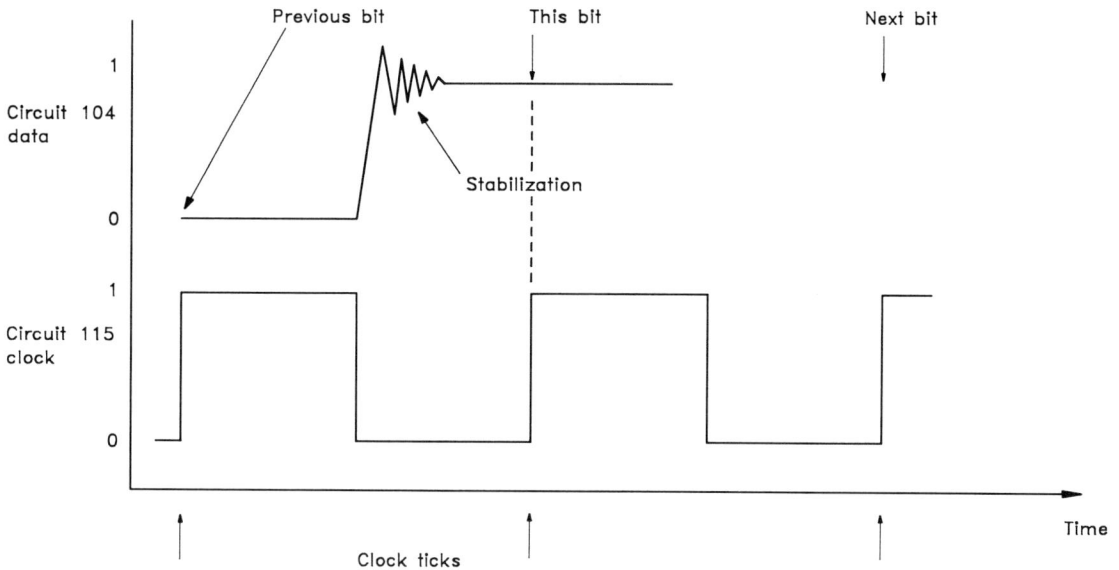

**Fig. 7.14** Receiver samples half a bit time after transmitter sets the line state; this allows line to stabilize

Clearly, the frequency of the one and zero cycle determines the frequency of the clock ticks and therefore the speed of the data transmission. A frequency of 2400 clock ticks per second would give a data rate of 2400 bps.

**Circuit 114**  *Transmitter Signal Element Timing DCE source*
(TSET-DCE)

This is the clock used for the Transmitted Data (circuit 103). It is used only when synchronous data is being communicated between the two devices.

The clock is generated by the DCE which means that the Transmitted Data and its clock have different sources and travel in different directions. The "Transmit" in the title refers to the data that the signal is clocking, not to the direction of this signal.

As with circuit 115, the clock has equal zero and one periods, and the zero-to-one transition marks the time at which the transmitter starts to set the circuit state. The DCE then samples the state of circuit 103 half a bit time later. It is important to appreciate that the DCE generates the clock and triggers the DTE to generate the data bit. The DCE then samples the bit half a bit time later.

Circuit 114 is the normal clock used for Transmitted Data; however, circuit 113 can be used as an alternative.

**Circuit 113**  *Transmitter Signal Element Timing DTE source*
(TSET-DTE)

This circuit provides a clock for the Transmitted Data (circuit 103), and is an alternative to circuit 114. The clock is generated by the DTE which means that the Transmitted Data and its clock both have the same source and travel in the same direction.

The choice of whether to use circuit 114 or 113 to clock the Transmitted Data is dependent on the capabilities of the particular DTE and DCE used. Clearly, if both components implement both circuits then the network manager can make a choice of which circuit to use, based upon the standards and practices of the network. Very often, one of the components will not implement one of the circuits and therefore dictates that the other must be used.

The network manager must check the capabilities of the equipment and connect one of the circuits. If neither is connected then there is no clock and no data can be transmitted by the DTE. If both circuits are connected then there will be two clocks and the results are not defined.

Circuit 113 follows the same general procedures as circuit 115. The clock tick marks the time at which the DCE should sample circuit 103, and the data bit is generated half a bit time earlier.

**Circuit 105**   *Request To Send* (RTS)
This is a circuit which, unlike clock and data lines that are bit-oriented, is either on or off. It is one of a number of control circuits whose state carries an indication from one end to the other of what is going on. The transitions between on and off are independent of other circuits, and result only from changes in what is happening in the equipment.

Circuit 105 is generated by the DTE and traditionally indicates that the DTE has data to transmit on circuit 103. The DCE responds on circuit 106 to indicate that the transmission can proceed.

This procedure is often unnecessary, for example in an X.25 network where the DCE is always ready to receive the Transmitted Data. In these circumstances the circuit is often used for hardware flow-control. If circuit 105 is on then the DTE can accept more data on circuit 104; if circuit 105 is off then the DCE cannot send any more Receive Data.

The use of circuit 105 for flow-control is very common in printers which must stop data arriving when the buffer gets full. The alternative is for the printer to generate an XOFF character on circuit 103, which is much more difficult to do. The subject of flow-control is also discussed in Chapter 2.

**Circuit 106**   *Clear To Send* (CTS)
              *Ready For Sending* (RFS)
This control circuit is the counterpart of circuit 105, and is generated by the DCE. Traditionally it indicates that the DTE may transmit data on circuit 103 and is used in response to a Request To Send on circuit 105.

It is more usually used for hardware flow-control, and indicates that the DCE is able to receive more data on circuit 103. If circuit 106 is off then the DTE cannot send any more Transmitted Data on circuit 103.

When circuits 105 and 106 are used for flow-control then they are completely independent of each other.

**Circuit 107**  *Data Set Ready* (DSR)

This is a control circuit generated by the DCE, indicating that the DCE is able to participate in data transfer. The DTE will therefore normally wait for this circuit to be on before it starts transmission.

The circuit is used primarily with modems on telephone circuits where there is a considerable amount of preliminary activity before data transfer can commence. Now that digital lines are readily available the telephone system is rarely used in high-speed communications networks.

Network managers are most likely to come across the circuit when a DTE requires the circuit but the DCE does not implement it. The DTE cannot then participate in data exchange. This type of problem is discussed later in this chapter.

**Circuit 109**  *Data Carrier Detect* (DCD)
              *Data Channel Received Line Signal Detector* (RLSD)

This control circuit is similar in function to circuit 107 and again indicates that the DCE is able to participate in data exchange. It is generated by the DCE.

Traditionally, it indicates that the carrier signal is present, which is a different state in the modem on a telephone network to merely being ready as indicated by circuit 107. On modern digital systems it simply shows that the DCE is operational and is the main way of determining whether the DCE is on-line and able to participate in data transfer.

**Circuit 108**  *Data Terminal Ready* (DTR)

This control circuit is the counterpart of circuit 109 and is generated by the DTE. It indicates that the DTE is operational and able to participate in data transfer.

V.24 actually defines two variants of the circuit which are traditionally used with modems on telephone circuits where the modem is able to receive incoming calls. When the modem detects the ringing signal for the incoming call it turns on circuit 125 (Ring Indicator) and must then gain authorization from the DTE to answer the call. This authorization is given via circuit 108.

In the first variant called Data Terminal Ready (DTR) or 108/2, the DTE turns on the circuit whenever it is able to operate. Thus DTR is on when the call arrives which means that the modem can answer the call immediately.

The second variant is 108/1 and is called Connect Data Set To Line (CDSTL). In this mode the DTE will turn on the circuit in response to the Ring Indicator. Thus the answering of each call must be authorized separately by the DTE.

Where the telephone network is not being used there is little distinction between the two and the circuit simply indicates that the DTE is on-line and ready.

**Circuit 111**  *Data Signalling Rate Selector DTE source* (SEL)

This control circuit is generated by the DTE, and is used to select the speed of transmission. For such a circuit to be useful then both the DTE and DCE must be capable of running at the same two speeds. The on state selects the higher speed.

The circuit is generally used with modems that have a low-speed backup connection to the other modem, perhaps via the dialled telephone network instead of a direct line. If the primary connection fails then the DTE can select the lower speed and the backup connection.

**Circuit 122**  *Secondary Data Carrier Detect* (SDCD)
*Backward Channel Received Line Signal Detector*

Where the application is two intelligent DTEs linked by two modems, then V.24 allows for two separate pairs of transmit and receive channels. This allows for the communication of error messages and supervisory information separately from the main data stream but simultaneously with it.

This system is very rarely used, and in modern networks is only likely to be useful for network management functions such as collecting statistics from a modem.

Circuit 122 is equivalent to circuit 109 for the secondary channel.

**Circuit 121**  *Secondary Clear To Send* (SCTS)
*Backward Channel Ready*

This is equivalent to circuit 106 for the secondary channel.

**Circuit 118**  *Secondary Transmitted Data* (SXMT)
*Transmitted Backward Channel Data*

This is equivalent to circuit 103 for the secondary channel.

**Circuit 119**  *Secondary Received Data* (SRCV)
*Received Backward Channel Data*

This is equivalent to circuit 104 for the secondary channel.

**Circuit 120**  *Secondary Request To Send* (SRTS)
*Transmit Backward Channel Line Signal*

This is equivalent to circuit 105 for the secondary channel.

**Circuit 125**  *Ring Indicator* (RI)
*Calling Indicator*

This control circuit is used with modems connected to telephone networks and indicates that an incoming call has arrived. It is not used on other networks.

**Circuit 141**  *Local Loopback* (L3)

This control circuit is generated by the DTE and causes the modem to enter the Local Loopback state. This allows the DTE to test the transmission and reception circuits and the modem itself, as explained in Chapter 3.

When the circuit is on, the modem enters the Local Loopback state and normal data transmission is not possible. The modem indicates to the DTE that it is in this state using circuit 142. It is not obligatory for modems to implement this circuit.

**Circuit 140** *Remote Loopback* (L2)
*Loopback/Maintenance test*

This control circuit is generated by the DTE and causes the local modem to enter Remote Loopback state. This in turn causes the local modem to signal to the remote modem, and causes that to enter digital loopback. This allows the local DTE to test the connection between the modems – the network – and the remote modem itself, as explained in Chapter 3.

When the circuit is on, the appropriate loopbacks are applied and normal data transmission is not possible. The modem indicates to the DTE that the loopbacks are applied using circuit 142. It is not obligatory for the modems to implement this circuit.

**Circuit 142** *Test Indicator* (TI)

This control circuit is generated by the DCE and indicates that a loopback condition exists on the local modem and that data transmission is not possible. The circuit is also turned on by a remote modem if a local modem requests remote loopback. The remote DTE is thus informed of the loopback requested by the local DTE.

## 7.9  ISO 2110

ISO 2110 defines the mechanical connection used to attach the DCE to the DTE. It also defines the V.24 circuits that will be used, and the correspondence between the pins of the connector and the various circuits.

The standard gives the precise physical dimensions of the twenty-five-way D-type connector that is commonplace on most items of data communications equipment.

- It assigns the male plug to the DTE.
- It assigns the female socket to the DCE.

This is illustrated in Fig. 7.15.

The plug and socket are then mated together to complete the circuits between the DTE and the DCE. If the DTE and DCE are not adjacent, then they can be connected using a cable that has a male plug at one end and a female socket at the other, and with all pins connected straight through.

Network components normally have the plug or socket directly mounted on the casing rather than on a flying cable, and a straight-

**Fig. 7.15** DTE on male plug, DCE on female socket

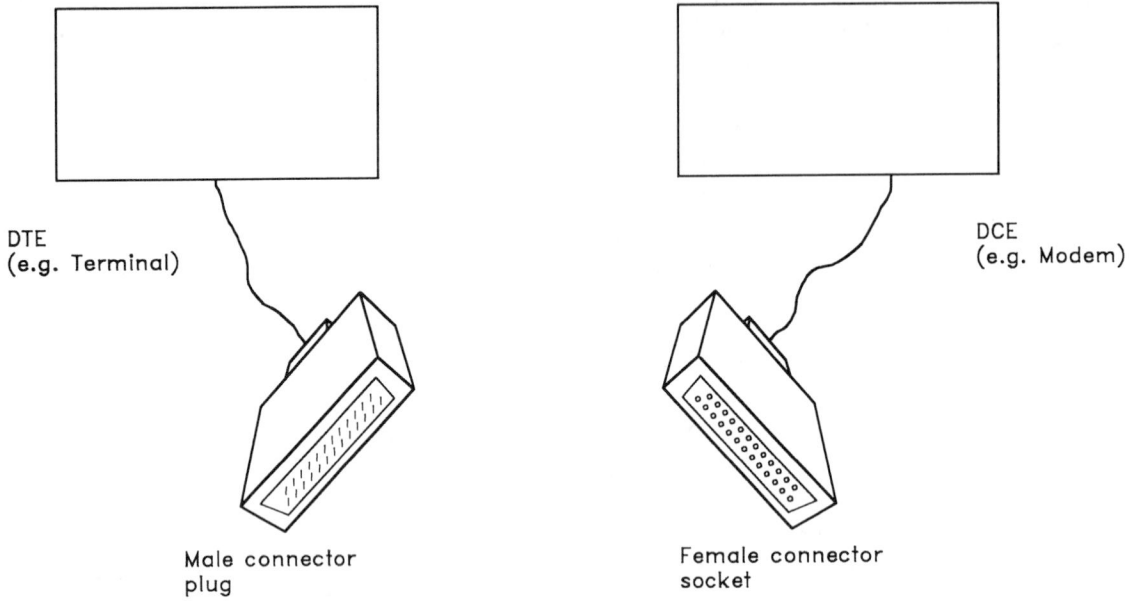

DTE
(e.g. Terminal)

DCE
(e.g. Modem)

Male connector plug

Female connector socket

through male-to-female cable is then necessary to make the connection.

The total length of the DTE-to-DCE connection, including any extension cables, is governed by the electrical specifications, normally V.28 with ISO 2110 connectors.

The assignment of connector pins to V.24 circuits is shown in Fig. 7.16. Pins 9, 10 and 11 are not assigned in ISO 2110. If any circuits are not needed in a particular DTE-to-DCE connection, for example the clock circuits in an asynchronous connection, then those circuits are not connected.

ISO 2110 defines a number of variations of pin assignments for specialized applications such as Telex, Automatic Calling Equipment, and for parallel data. The assignment shown in Fig. 7.16 is the most common for data communications equipment.

**Fig. 7.16** Assignment of V.24 circuits to ISO 2110 connector

| ISO 2110 pin number | V.24 circuit | | | Direction DTE DCE |
| --- | --- | --- | --- | --- |
| 1 | | Protective Ground | (PGND) | |
| 2 | 103 | Transmitted Data | (XMT) | → |
| 3 | 104 | Received Data | (RCV) | ← |
| 4 | 105 | Request to Send | (RTS) | → |
| 5 | 106 | Clear to Send | (CTS) | ← |
| 6 | 107 | Data Set Ready | (DSR) | ← |
| 7 | 102 | Signal Ground | (GND) | |
| 8 | 109 | Data Carrier Detect | (DCD) | ← |
| 9 | | | | |
| 10 | | | | |
| 11 | | | | |
| 12 | 122 | Secondary Data Carrier Detect | (SDCD) | ← |
| 13 | 121 | Secondary Clear to Send | (SCTS) | ← |
| 14 | 118 | Secondary Transmitted Data | (SXMT) | → |
| 15 | 114 | Trans. Signal Element Timing | (TSET–DCE) | ← |
| 16 | 119 | Secondary Received Data | (SRCV) | ← |
| 17 | 115 | Recv. Signal Element Timing | (RSET) | ← |
| 18 | 141 | Local Loopback | (L3) | → |
| 19 | 120 | Secondary Request to Send | (SRTS) | → |
| 20 | 108 | Data Terminal Ready | (DTR) | → |
| 21 | 140 | Remote Loopback | (L2) | → |
| 22 | 125 | Ring Indicator | (RI) | ← |
| 23 | 111 | Data Signalling Rate Selector | (SEL) | → |
| 24 | 113 | Trans. Signal Element Timing | (TSET–DTE) | → |
| 25 | 142 | Test Indicator | (TI) | ← |

## 7.10  V.28

This recommendation defines the electrical characteristics of the DTE-to-DCE circuits. The circuit runs between what the recommendation calls a generator at one end and a load at the other. The generator and load pair are independent for each V.24 circuit, and may be located either way round in the DTE and DCE.

V.28 introduces the concept of the *Interchange Point*, which is the line of demarcation between the DCE and DTE. The Interchange Point is located at the connectors between the DCE on one side and the DTE and any cables on the other. This is shown in Fig. 7.17.

**Fig. 7.17**  Position of Interchange Point in DTE-to-DCE connection

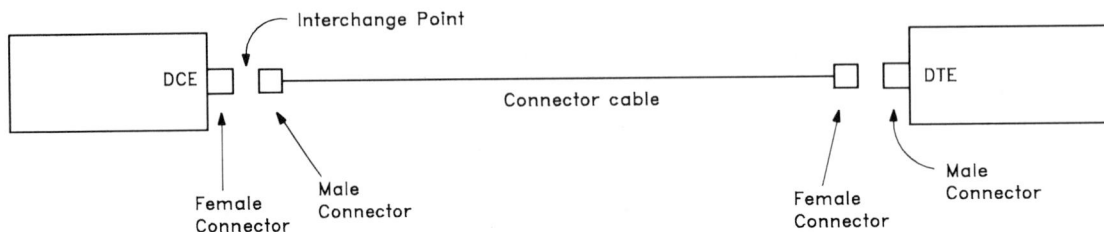

**Fig. 7.17**  Position of Interchange Point in DTE-to-DCE connection

As shown in section 7.8 the potential difference on the circuits is measured with respect to Signal Ground (circuit 102). The Interchange Point is the place at which the measurement is taken. This is, of course, theoretical since the load — and therefore the detector components — are located within the DCE or DTE. The recommendation therefore defines the characteristics of the circuit, so that the detectors can assess the remote potential difference at the Interchange Point.

The potential difference between any circuit and Signal Ground at the Interchange Point, has the following significance:

- If the potential difference is greater (more positive) than three volts:
  *a*) For data and clock lines the circuit is in the ZERO state
  *b*) For control lines the circuit is in the ON state.
- If the potential difference is less (more negative) than three volts:
  *a*) For data and clock lines the circuit is in the ONE state.
  *b*) For control lines the circuit is in the OFF state.

The size of the potential difference is not significant, as long as it is bigger than three volts — it is the polarity that is important.

If the potential difference is between positive three volts and negative three volts then the line state is indeterminate. The indeterminate state can only occur when the circuit is in transition from one to zero or zero to one, and the recommendation states that the transition must take less than one millisecond. Figure 7.18 shows some examples of circuit voltages and corresponding line states.

V. 28 states that the generator should create a potential difference of between three and fifteen volts, of either polarity, at the Interchange Point, unless in the transition state. Most manufacturers make generators that always produce twelve volts at the generator output, and then rely on the circuit characteristics to ensure that the voltage is correct at the Interchange Point.

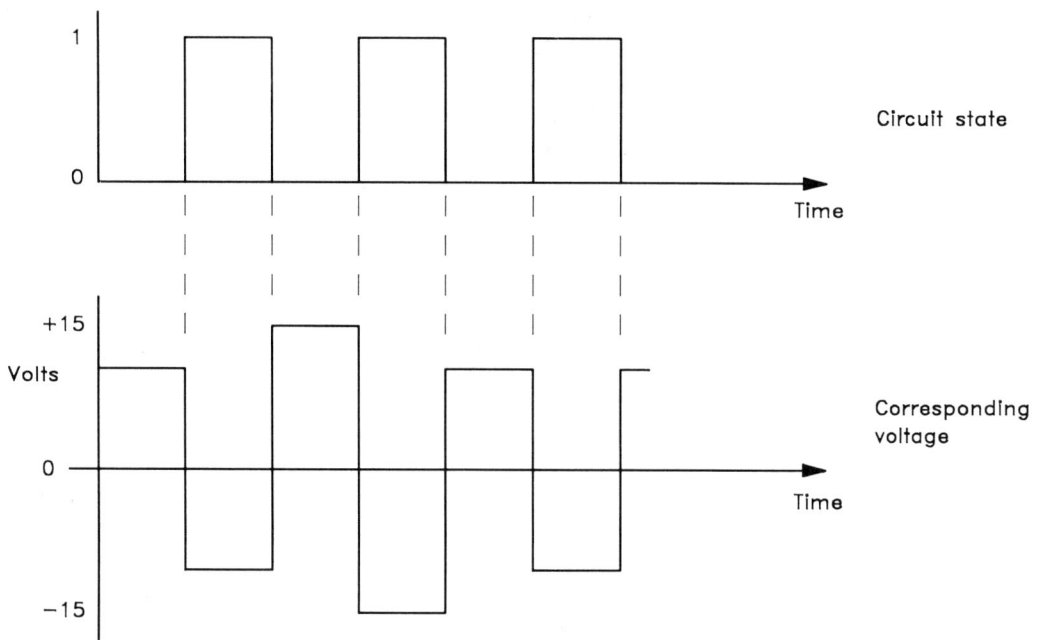

CIRCUIT 114    TRANSMITTER SIGNAL ELEMENT TIMING

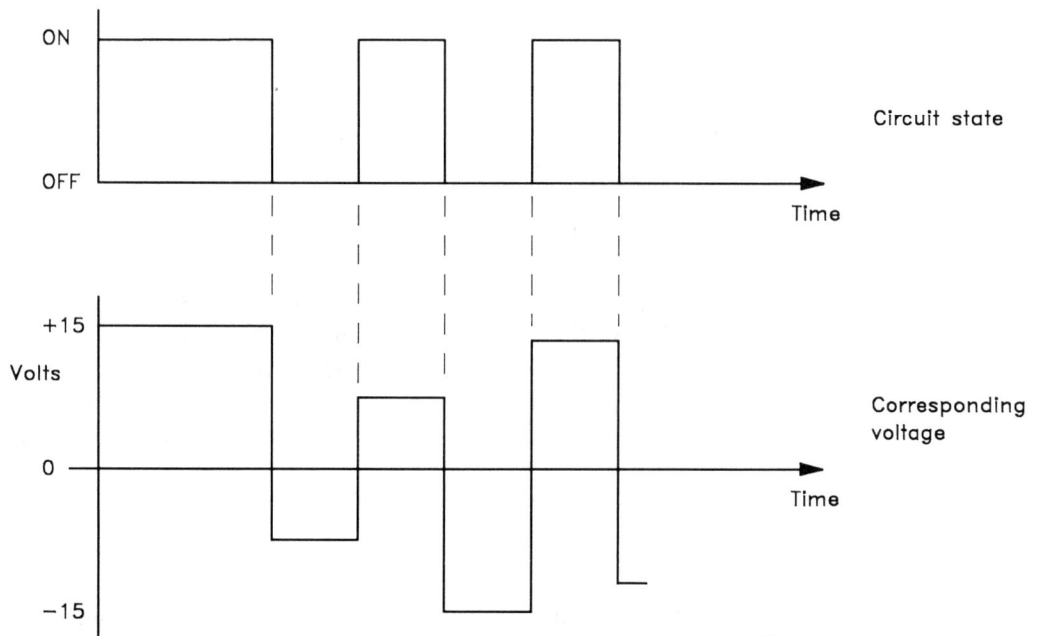

CIRCUIT 108    DATA TERMINAL READY

**Fig. 7.18** Voltages and corresponding line states; note that value of voltages is not significant above three volts; transition time must be less than one millisecond

Most people will have experienced the phenomenon of a short burst of interference on a television or radio receiver when a nearby light is turned on. This is due to electromagnetic radiation generated by the sudden burst of electricity flowing to the light. Data communications equipment is also susceptible to electromagnetic radiation, which can corrupt the data by inducing a potential difference on the circuits. This normally occurs in one of three ways:

● Radiation through the air which is picked up by the equipment directly, or picked up by the cables between the equipment. The cables can be protected to a large extent by using a screen which is wired to Protective Ground (see section 7.8). The screen is a metal braid or foil wrapped around the conductors of the circuits, which forms a barrier to the radiation. The screen is sometimes said to form a Faraday Cage.

● Surges in electrical supply cables can be caused by nearby equipment being turned on or off. These surges can then be carried inside the equipment by the mains electrical connection. These surges are relatively easy to filter out in the power supply of the equipment. If they are not filtered then they can travel through the equipment and into the conductors of the data circuits. They are then able to travel into other items of equipment and affect them, even though the original equipment was able to sustain the surge without problem.

● Surges can be caused in the conductors of a data circuit, by nearby conductors switching between zero and one or on and off.

The latter problem is an obvious cause of concern in DCE-to-DTE cables where many circuits are laid close together, and all are capable of a transition in state at any time. To reduce the possibility of false signals being induced in this way, V.28 limits the rate at which signals may change state, and therefore reduces the surges.

The fastest transition allowed is at a rate of thirty volts per microsecond, and thus the transition must be sloped as shown in Fig. 7.19. This slope is also known as the *slew rate*, and the circuit is said to be slew rate limited. Slew rate limiting must be balanced against the requirement of a transition time of less than one millisecond.

## 7.11 Cable length

V.28 does not define the length of cable allowed between the DCE and DTE, it merely defines the electrical characteristics of the cable. The following problems grow in severity as the cable length increases:

● *Voltage drop* As electrical energy travels around the circuit, some is dissipated, most of it in the form of heat. Therefore a potential difference of three volts at the Interchange Point will

+15

30 Volts

One microsecond

Zero or ON

Volts

0

Time

Slope must be
flatter than this

−15

One or OFF

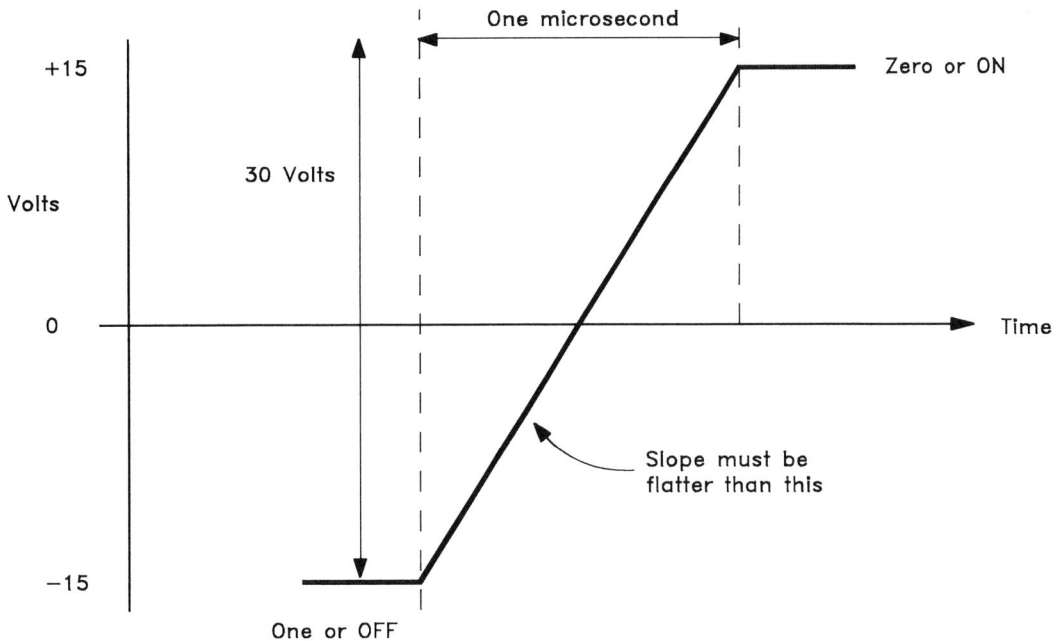

**Fig. 7.19** Slew rate
limiting to reduce induction

appear to be much less at the end of a long cable. This effect can be
reduced by using thicker cable that has less electrical resistance.

● *Radiation*   The cable gets more and more like an aerial as its
length increases, and picks up significantly more radiated energy.
This effect can be reduced by using higher quality screening on the
cable.

● *Induction/Capacitance*   As the length increases, the cable no
longer passes a true signal. In particular, the transition slopes are
modified and may cease to conform to the recommendation even
though they were generated correctly.

It is rather like a radio receiver being moved further and further from
the transmitter: eventually the receiver becomes unable to distinguish
the real signal from all the false signals.

The reason that V.28 is not explicit about length is that it is not
possible to be precise. There are so many factors affecting transmission,
most of which are invisible and difficult to measure. It is only possible to
be pessimistic and quote short lengths, in the full knowledge that longer
lengths are usually possible depending on the actual installation.

There are test instruments that try to determine the quality of the line,
but the only truly practical way of determining whether it is too long is
to run it with the DTE and DCE that will use it. The error rate should
then be carefully monitored by whatever means are available. This
might be done by looking at the number of Reject frames on an X.25
line, or by gathering statistics from the application.

If there is a problem and the error rate is not acceptable, then line
drivers or modems will have to be used.

## 7.12    Some example connections

**Fig. 7.20**  Terminal
connection on four-wire
circuit

**Example 1**  Figure 7.20 shows a terminal connected to a computer using asynchronous full-duplex transmission. This is pretty well the simplest possible connection. The circuits connected are Signal Ground (GND) (circuit 102), Transmitted Data (XMT) (circuit 103), Receive Data (RCV) (circuit 104), and Protective Ground (PGND).

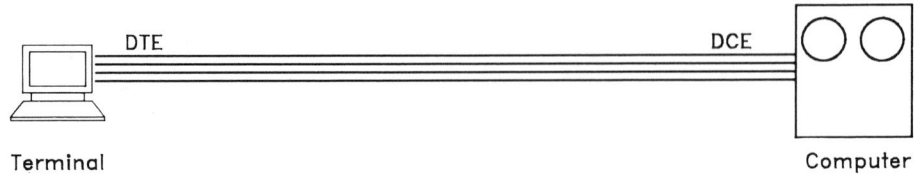

Terminal                                                            Computer

**Fig. 7.21**  Stage one
drawing for simple terminal
connection

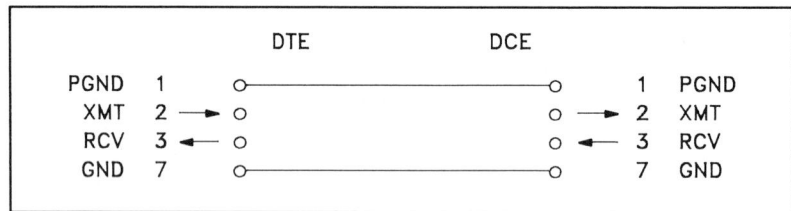

**Fig. 7.22**  Stage two
drawing for simple terminal
connection

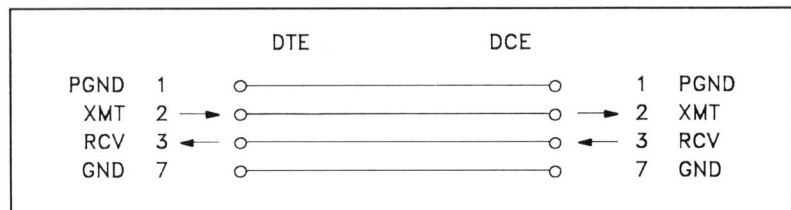

When designing connections it is helpful to adopt a two-stage approach. Firstly, the DTE and DCE connectors are drawn with the circuits that are implemented by the devices. On this drawing Signal Ground and Protective Ground are drawn between the two ends since they are needed in every connection. This first-stage drawing is shown in Fig. 7.21 for the simple terminal.

The second stage is to draw the remaining circuits, taking account of the individual capabilities of the two ends. In the case of the terminal this is fairly trivial and the complete drawing is shown in Fig. 7.22. From this drawing the connecting cable can be constructed.

**Example 2**  A synchronous device with DTE connector is to be connected to a modem. Data will be synchronous and the transmit clock will be provided by the modem on circuit 114, Transmitter Signal Element Timing DCE source (TSET-DCE). It is necessary for this to be specified since there are two possible clocks for the Transmitted Data as discussed earlier. The device will assert Data Terminal Ready (DTR) (circuit 108) when it is operational. The modem uses flow control with

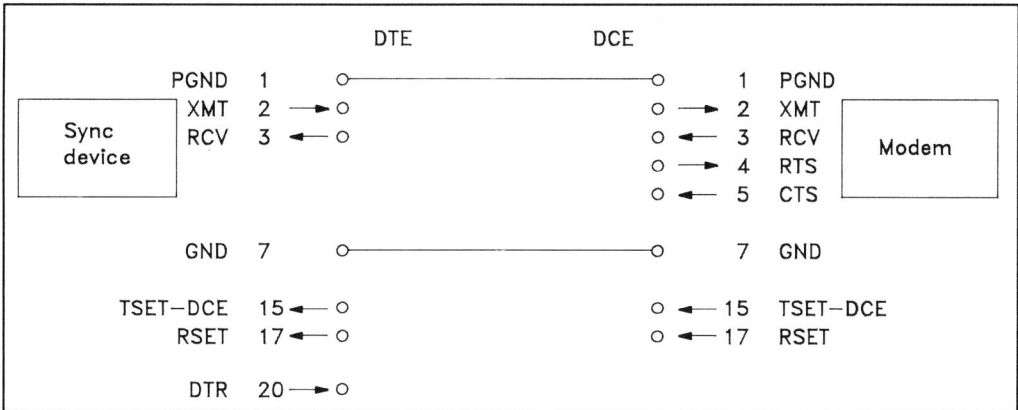

**Fig. 7.23** Stage one drawing for sync connection to modem

Ready To Send (RTS) and Clear To Send (CTS) (circuits 105 and 106).

The first-stage drawing is shown in Fig. 7.23. Some of the circuits are easy to complete since they are implemented by both ends. The data circuits and clock circuits can just be drawn in end-to-end.

The two flow control circuits of the DCE are intended to indicate when the relevant end can accept more data. The DCE can accept data as long as it keeps CTS on, and the DCE will deliver data as long as RTS in on.

The RTS circuit ought to be connected to the DTE to allow it to control the flow of Received Data, but in this configuration the DTE does not supply the circuit. The circuit must be kept on or the DCE will not supply any Received Data. To make the configuration work, the RTS circuit must be connected to the DTR circuit of the DTE. Although this will not control the flow of data, it will allow the data to flow.

There is nothing that can be done with CTS since the DTE does not implement any way of controlling Transmitted Data. Had the DTE not implemented DTR, then the RTS and CTS of the DCE could have been connected together to allow data to flow.

None of these solutions is ideal because they do not protect against overflow due to data being sent too quickly. They are nevertheless practical solutions to the problems. The stage-two drawing is shown in Fig. 7.24.

**Fig. 7.24** Stage two drawing for sync connection to modem

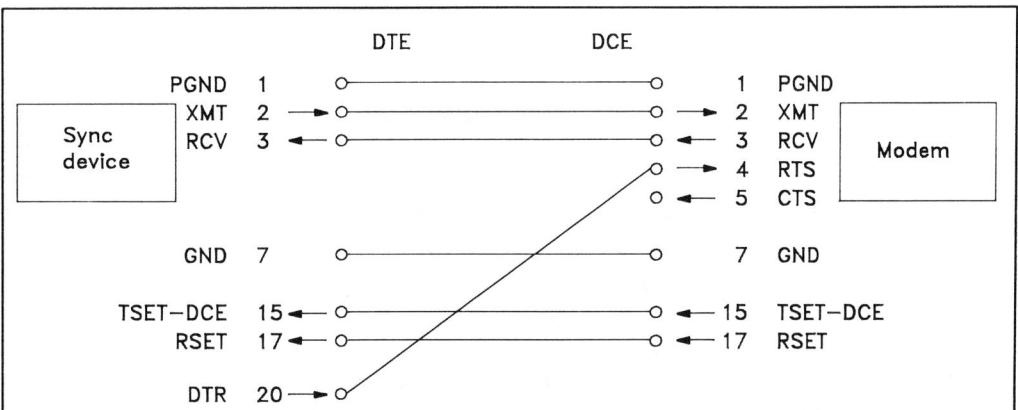

**Example 3** Suppose that in a classroom situation it is necessary to connect two terminals to each other, so that characters typed on one appear on the screen of the other and vice-versa. Both terminals will normally be DTEs and hence cannot be connected directly. However, drawing the first diagram shows a solution. This is shown in Fig. 7.25. Note that both terminals implement flow control using RTS and CTS.

**Fig. 7.25** Stage one drawing for connecting two terminals

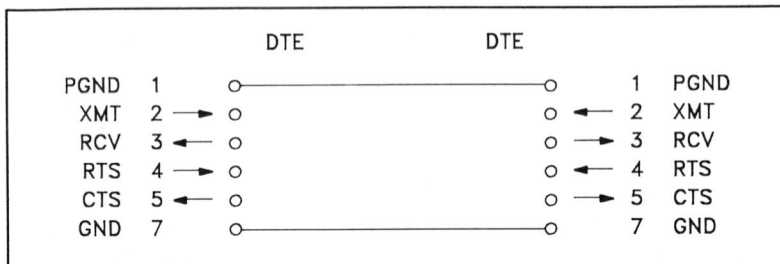

**Fig. 7.26** Stage two drawing showing Null Modem

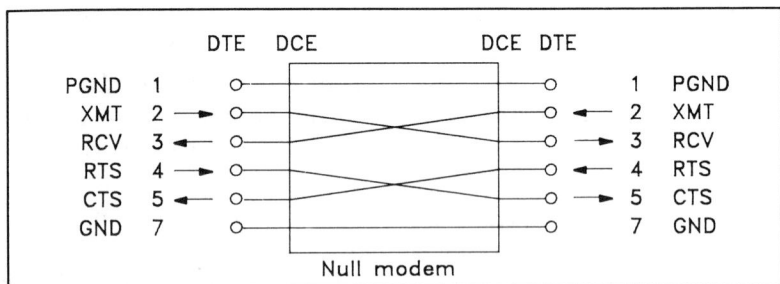

The arrows on the circuits show that interconnection is possible because for every generator there is a load. The connection is achieved by crossing the circuits over in a device called a Null Modem. This is shown in Fig. 7.26.

The Null Modem is so called because it provides two DCE connectors to connect two DTEs together. It therefore takes the place of two modems and the intervening network. The precise circuits used in the Null Modem will depend on the capabilities of the two DTEs.

Rather than using two cables from the DTEs to the separate Null Modem, the whole connection can be achieved in a single cable that performs the appropriate crossovers.

A similar mechanism can be used to connect two asynchronous DCEs.

**Example 4** An X.25 link is to be implemented between a PAD and an adjacent Switch. The PAD has a DTE interface and the Switch a DCE. The PAD implements TSET-DTE (circuit 113), the DTE transmit clock, and can also be configured to receive TSET-DCE (circuit 114), the DCE clock. The Switch only implements TSET-DCE (circuit 114) so the PAD will have to be configured for this circuit. Both units implement RSET (circuit 115), the Receive clock. Both the PAD and the Switch implement DTR, Data Terminal Ready (circuit 108), and DCD, Data Carrier Detect (circuit 109), as an indication that they are operational. The PAD

**Fig. 7.27** diagram:

```
                    DTE            DCE
PGND       1    o----------------o    1   PGND
XMT        2 --> o               o --> 2   XMT
RCV        3 <-- o               o <-- 3   RCV

DSR        6 <-- o

GND        7    o----------------o        7   GND

DCD        8 <-- o               o <-- 8   DCD

TSET-DCE  15 <-- o               o <-- 15  TSET-DCE
RSET      17 <-- o               o <-- 17  RSET

DTR       20 --> o               o --> 20  DTR

TSET-DTE  24 --> o
```

[PAD]                                    [Switch]

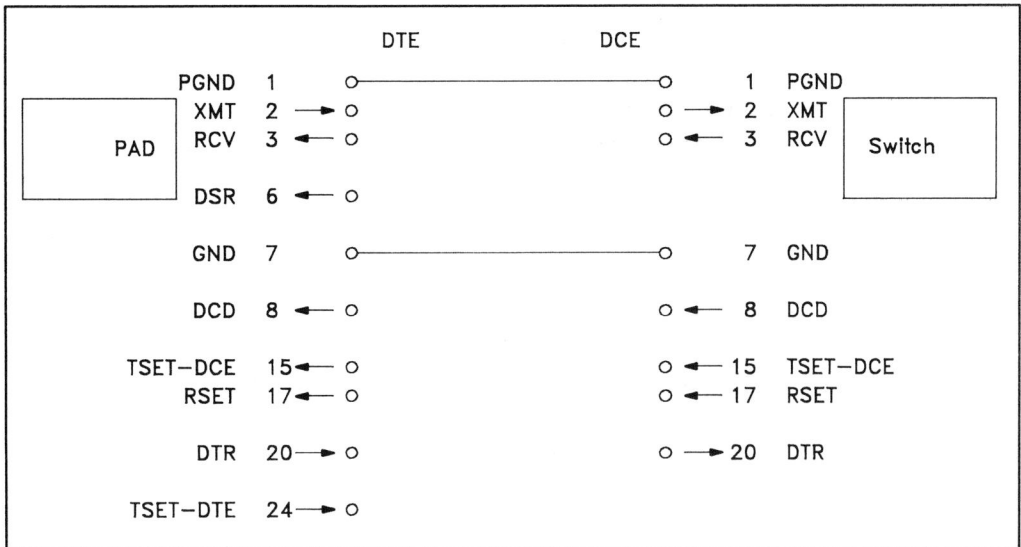

**Fig. 7.27** Stage one drawing for PAD to switch connection

**Fig. 7.28** Stage two drawing for PAD to switch connection

also implements DSR, Data Set Ready (circuit 107), and must have this circuit on before it will become operational and raise DTR.

The first drawing is shown in Fig. 7.27. The only complexity in this configuration is the supply of an operational indication to DSR of the PAD. In some items of equipment this could be connected to DTR so that the PAD persuades itself that the other end is operational. This is feasible in general, but in this case will not work because the PAD will not raise DTR until DSR has ben raised. Instead, DSR must be connected to DCD, so that the DCE raises both circuits as soon as it becomes operational.

There is generally no problem in connecting a generator to several loads, though more than say five would be imposing on good fortune.

Figure 7.28 shows the completed drawing using DCD to hold DSR on.

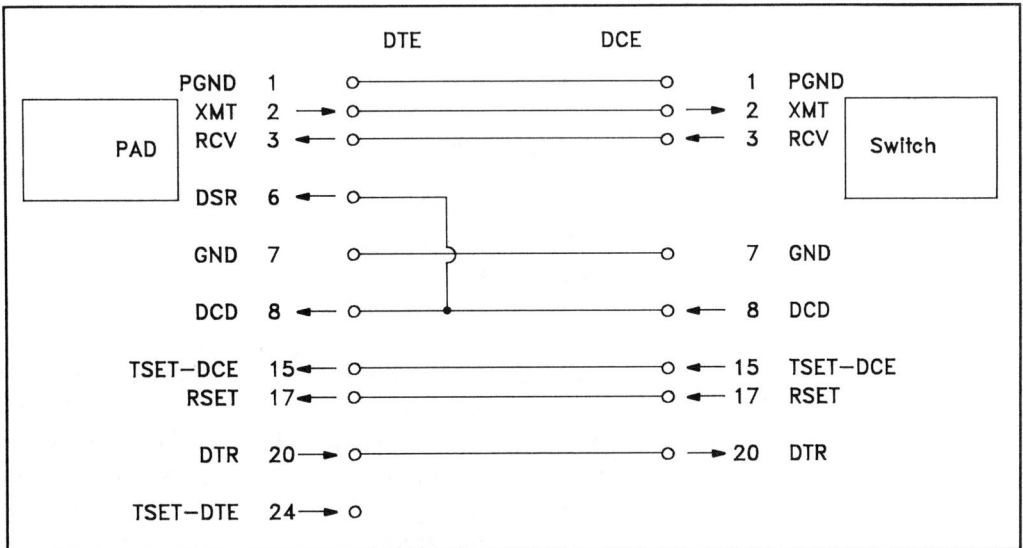

**Fig. 7.28** diagram:

```
                    DTE            DCE
PGND       1    o----------------o    1   PGND
XMT        2 --> o---------------o --> 2   XMT
RCV        3 <-- o---------------o <-- 3   RCV

DSR        6 <-- o----+
GND        7    o-----+----------o        7   GND
DCD        8 <-- o----+----------o <-- 8   DCD

TSET-DCE  15 <-- o---------------o <-- 15  TSET-DCE
RSET      17 <-- o---------------o <-- 17  RSET

DTR       20 --> o---------------o --> 20  DTR

TSET-DTE  24 --> o
```

[PAD]                                    [Switch]

195

**Example 5** A printer with a DTE interface is to be connected to the asynchronous DTE interface of a computer. The printer implements DSR, Data Set Ready (circuit 107), and requires the computer to indicate that it is operational using this circuit before the printer will become operational. The computer does not implement DSR. The printer has a limited buffer size and uses DTR, Data Terminal Ready (circuit 108), to indicate that the buffer is full – DTR is on when it can accept data. The computer implements flow control on RTS and CTS (Ready To Send and Clear To Send).

The use of control lines in this individual way is unfortunately commonplace, and network managers should never assume that manufacturers share their enthusiasm for standardization.

Figure 7.29 shows the completed diagram for this example.

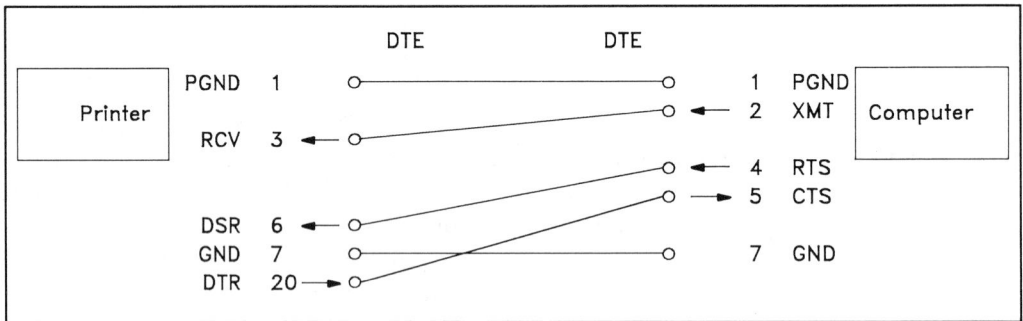

**Fig. 7.29** Stage two drawing for printer to computer connection

The use by the printer of DTR matches the normal flow control usage, so this can be connected to CTS on the computer.

The computer holds RTS on whenever it can accept data. Because no data is sent from the printer, RTS will be on as long as the computer is operational, and this can therefore be connected to DSR on the printer. If the computer implemented DTR on pin 20, this could have been used instead and would have been more aesthetically pleasing.

**Example 6** Example 4 showed a synchronous link between a PAD and a Switch that was a convenient DCE-to-DTE connection. If this had been two DTEs instead then would the connection have been possible by using a Null Modem arrangement?

There is no problem with the data and control circuits, but the clock circuits present more difficulty. Figure 7.30 shows the first-stage drawing for the data and clock circuits.

As long as both DTEs can generate the transmit clock (TSET-DTE), then the connection can be made by crossing both TSET-DTEs with Receive Signal Element Timing (RSET) as shown in Figure 7.31. The diagram shows the clock that is associated with each of the data circuits.

If one of the DTEs cannot generate TSET-DTE then the connection is still possible as shown in Figure 7.32. Here the single clock generator is connected to three clock loads. The data generated by DTE-A is clocked as before, but data generated by DTE-B is being clocked incorrectly.

**196**

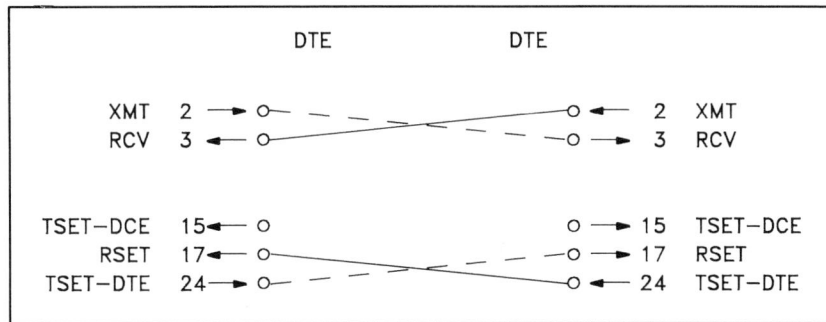

**Fig. 7.30** Connecting synchronous DTEs

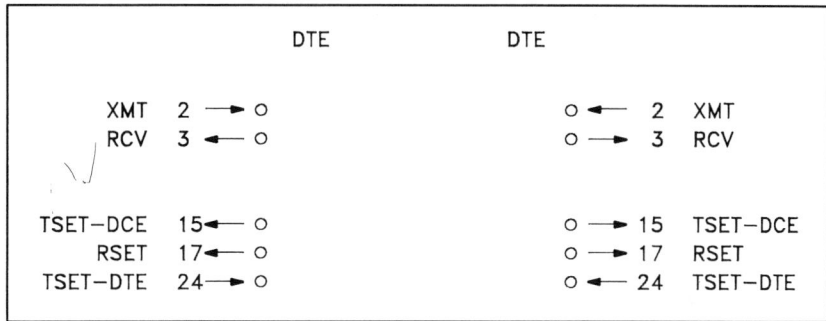

**Fig. 7.31** Corresponding data and clock circuits

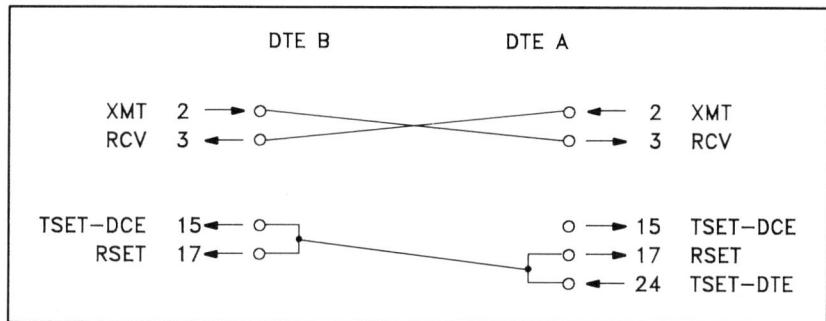

**Fig. 7.32** Transmitter and receiver clocks from same source

As far as DTE-B is concerned, it is receiving TSET-DCE and, as such, starts to set the state of the data circuit on the clock tick. As far as DTE-A is concerned, it is receiving RSET and, as such, checks the state of the data circuit on the clock tick. This means that transmission and reception of the data bits transmitted by DTE-B occurs simultaneously, rather than half a bit time out of synchronization as is normal. There is therefore no chance for the state of the data circuit to stabilize between transmission and reception, and errors are likely to be introduced.

Figure 7.33 summarizes the clock synchronizations, and illustrates why the problem described above comes about.

If neither DTE can be configured to generate TSET-DTE then there is no clock available and direct connection between the DTEs is not possible.

A variant of the Null Modem, called a Modem Eliminator, is available for this type of application. This is a device which generates clocks internally.

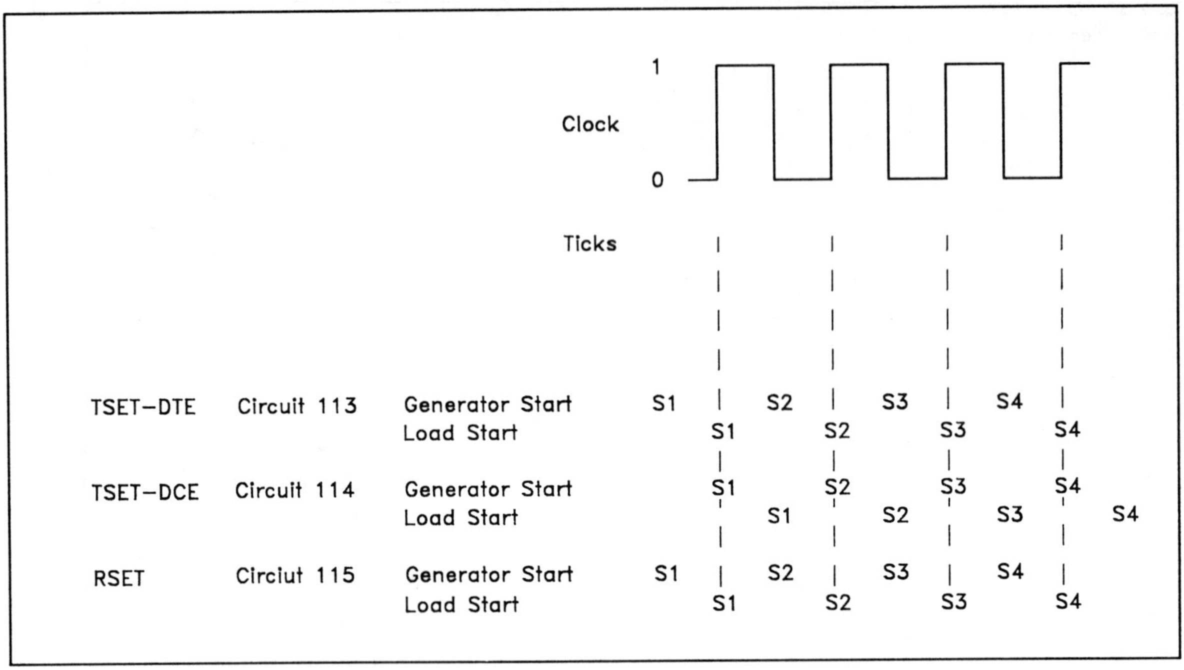

**Fig. 7.33** Clock synchronizations

**Example 7** Two X.25 physical DCEs are to be connected. Both can be configured for TSET-DTE (circuit 113) as well as TSET-DCE (circuit 114) for the source of the transmit clock. Both use flow control with RTS, Ready To Send, and CTS, Clear To Send, circuits. Both generate DCD, Data Carrier Detect, and require DTR, Date Terminal Ready, to be on before data can be transferred. Additionally, they exchange network management data asynchronously on the secondary channel. Figure 7.34 shows the diagram for this connection.

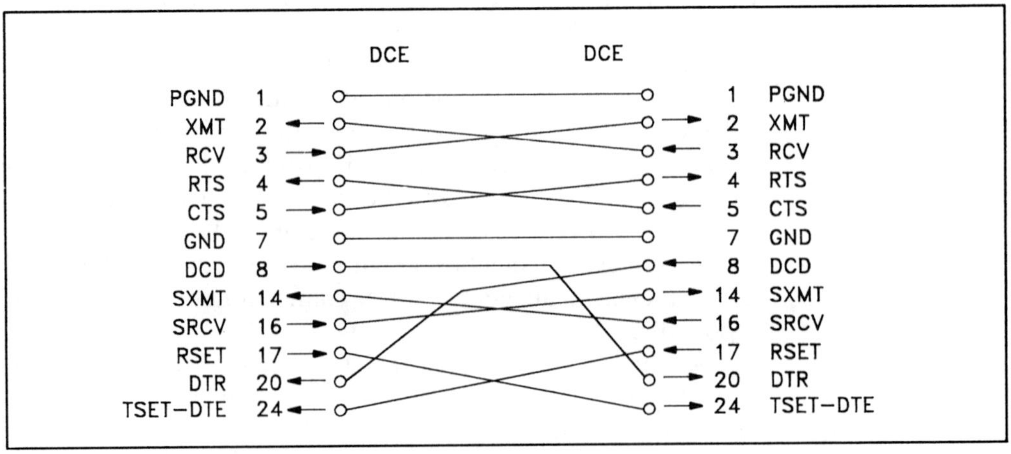

**Fig. 7.34** Two DCE's connected using secondary channel

## 7.13  Constructing the cable

The quality of cable construction is a significant factor in the quality of signals, and therefore on the number of errors generated in the circuits. If proper attention is given to the construction then cable runs can be longer before line drivers or modems become necessary. The following items can improve a cable:

- The conductors should be thick so as to be mechanically resilient and so as to present less electrical resistance. Typical conductors would have a cross-sectional area of 0.2 − 0.4 square millimetres.
- The cable should have the minimum number of conductors necessary for the application − extra conductors only exacerbate the radiation problems. If one or two extra conductors are unavoidable they should be connected to Signal Ground.
- Conductors must be connected to a generator at one end and a load at the other. The worst thing is to connect just one end since this will act as an aerial, and either generate radiation to other conductors, or pick up radiation from them.
- Ribbon cable is not a good idea unless conditions are ideal. The conductors are thin and unscreened, and all twenty five circuits are wired through.
- The quality of screening determines the susceptibility to electro-magnetic radiation and ranges from a simple foil around all the conductors, through separate foils around groups of conductors, to braided shields. Multiple screening is also available offering even more protection.
- The conductors must be grouped according to the electrical standards. V.24/ISO 2110/V.28 connections should have conductors that are not twisted around each other. This minimizes the induction of signals from one to the other. X.21/V.11 cables are required to have pairs of conductors twisted together.
- If the conductors are soldered to the connectors then the joint must be sound. Poor joints, although undetectable to the naked eye, have increased electrical resistance and may be intermittent, thus corrupting the data on the circuit. Connectors with crimp pins are easily available and give a more consistent quality.
- Many qualities of connectors are available that meet the ISO 2110 or other physical specifications. An "economical" connector may have a limited life or may mate poorly with its counterpart.
- When the cable enters the hood of the connector the screening has to be stripped back to allow the conductors to be connected to the pins. This leaves a few centimetres of conductor unshielded. The shielding can be restored by using a metal hood for the connector, or a metalized plastic one. The metal hood, and the metalwork of the connector, must then be attached to Protective Ground to ensure that the entire length of the cable is protected. Whatever

hood is used it should have an efficient strain relief to anchor the cable, and prevent the weight of the cable pulling the conductors off the pins of the connector.

- There is no point going to the expense of a good cable if it keeps falling out the back of the computer. It is therefore necessary to ensure that appropriate locking devices are fitted. Many types are available offering features such as quick release, small dimensions, and security from prying fingers
- It is helpful if the cable is suitably marked to identify its function. This reduces the possibility of disconnection by mistake.
- The cable should be routed as far away from other electrical equipment as possible. Where it is laid in ducts or trunking there should be no mains wiring in the same duct or trunk. Where the cable is exposed, it should be protected against physical damage such as people walking on it, and its installation should protect the people from tripping over it.

There is some debate as to whether the screen should be connected at both ends of the cable. The reason for doing this is to preserve what is called the Faraday Cage and to provide protection against induced signals throughout the length of the cable. The disadvantage is that it may compromise the electrical safety of the installation, especially where the two ends of the cable are located in different buildings with different electrical earths. In general, long cable runs will use modems, and therefore most data cables that a network manager might construct will be local and the electrical considerations probably will not apply.

## 7.14 The breakout box

The documentation provided with a particular interface is often not sufficiently clear to allow a cable to be designed and constructed with complete confidence. Often, some sort of experimentation is necessary, and it is always desirable to avoid constructing a cable that is not correct.

The Breakout Box is the generic name given to a variety of pro-prietary devices that enable experiments to be performed. The devices are generally the size of a pack of playing cards and have connectors that allow the box to be inserted into a connection between two pieces of equipment. Normally, a pin-to-pin cable with all twenty five circuits wired is required to make-up the distance.

Most boxes have a number of lights on them which show the state of the circuits. They also have a bank of switches allowing the twenty five circuit paths to be broken. With just these features it is usually possible to see from the lights what is being generated, and to disconnect the unnecessary circuits.

Where the boxes really come into their own is when they are provided with a patching capability. The twenty five circuits on both sides of the isolation switches are made available on the control panel, and wires are provided to allow connections from any circuit to any other.

Clock circuits are a difficulty with Breakout Boxes because they oscillate too quickly for the lights to come on. Many boxes therefore feature a "pulse extender" which allows the oscillation to be seen.

Some boxes have lights which can show two colours thus allowing the state of the circuit to be seen: red = one/off, green = zero/on, both off = transition. Simpler boxes have a single colour which can only show limited information.

The following example shows the use of a Breakout Box to create a connection between asynchronous devices of unknown capabilities and requirements. Suppose a printer is being attached to a computer as shown in Fig. 7.35.

**Fig. 7.35** Breakout Box inserted in DTE-to-DCE connection

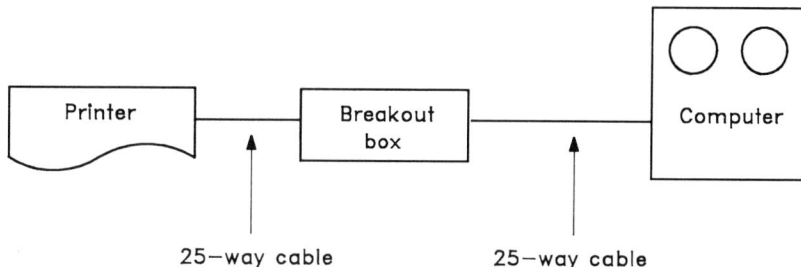

The first thing to do is to adjust the switches on the box so that all circuits are disconnected except for Signal Ground and Protective Ground. Pin 2 of the computer side is then patched to the pulse extender. Fig. 7.36 shows what the box may look like.

With pin 2 connected to the pulse extender, we make the computer output some data. If the light on the extender now comes on, then we know that the computer outputs on pin 2 so is probably wired as a DTE. If the light does not come on, the extender can be connected to pin 3 of the computer to check that it is a DCE.

If the extender light does not indicate data in either case then the computer needs some of its other circuits to be held in an on state. The box will have an "on" voltage available which should be connected as follows:

- Firstly, to pins 5, 6 and 8 which in the case of a DTE are inputs to CTS, DSR, and DCD loads. The extender is connected to pin 2 which is XMT output from a DTE.
- If no data is observed then connect the "on" to pins 4 and 20 which in the case of a DCE are inputs to RTS and DTR load. The extender is connected to pin 3 which is RCV output for a DCE.

One of these should have resulted in data being observed on the extender. Suppose it shows the computer to be a DTE, so that XMT data is output on pin 2. We can now try to connect the signals across to the printer.

- Firstly assume the printer to be a DCE. Although this is unlikely it is easier to try this because the circuits are straight through and we can use the switches on the box. The "on" from the box is still connected to CTS, DSR and DCD of the computer. Turn on the switch for pin 2 – this connects the XMT output of the computer to the XMT input of the DCE. Data can now be sent from the computer and be observed on the extender light which is still connected, and should be printed. If it is not printed then this may be because some control circuits are needed by the printer. Connect the "on" voltage to pins 4 and 20 which are RTS and DTR loads for a DCE. If still nothing is printed than the printer is likely to be a DTE.
- To try a DTE, first of all turn off the switch for pin 2. Pin 2 of the computer, which is XMT output, should be patched across to pin 3 of the printer, which is RCV input for a DTE. If nothing is printed then connect the "on" voltage to the control circuits by turning on the switches for pins 5, 6 and 8. These pins should still be connected to the "on" source on the computer side of the box. The printer should now work when data is sent by the computer.

We have now established that both items of equipment are DTEs, and that some combination of the control circuits on both sides needs to be kept on. This combination should now be explored.

Firstly remove the "on" from pin 8 leaving it connected to pins 5 and 6, and instead connect an "off" source to pin 8.

- If everything still works then neither side requires DCD to be on. The pin 8 switch should be turned off and the circuit can be ignored.
- If the printer stops then one or both sides require DCD.

    Firstly, turn off the switch for pin 8, leave the "off" connected to the computer side, and connect "on" to the printer side. If everything works then the printer needs DCD and the computer does not.

    Secondly, reverse the "off" and "on" connections so that the "on" is on the computer side. If everything works then the computer needs DCD and the printer does not.

    Thirdly, remove the "off" connection and turn the switch back on, so that "on" is connected to both sides again. This should work, since it is what we had originally, and shows that both the computer and the printer need DCD.

This process should be repeated for pins 5 and 6 to determine the requirements of each side. Suppose this shows the following:

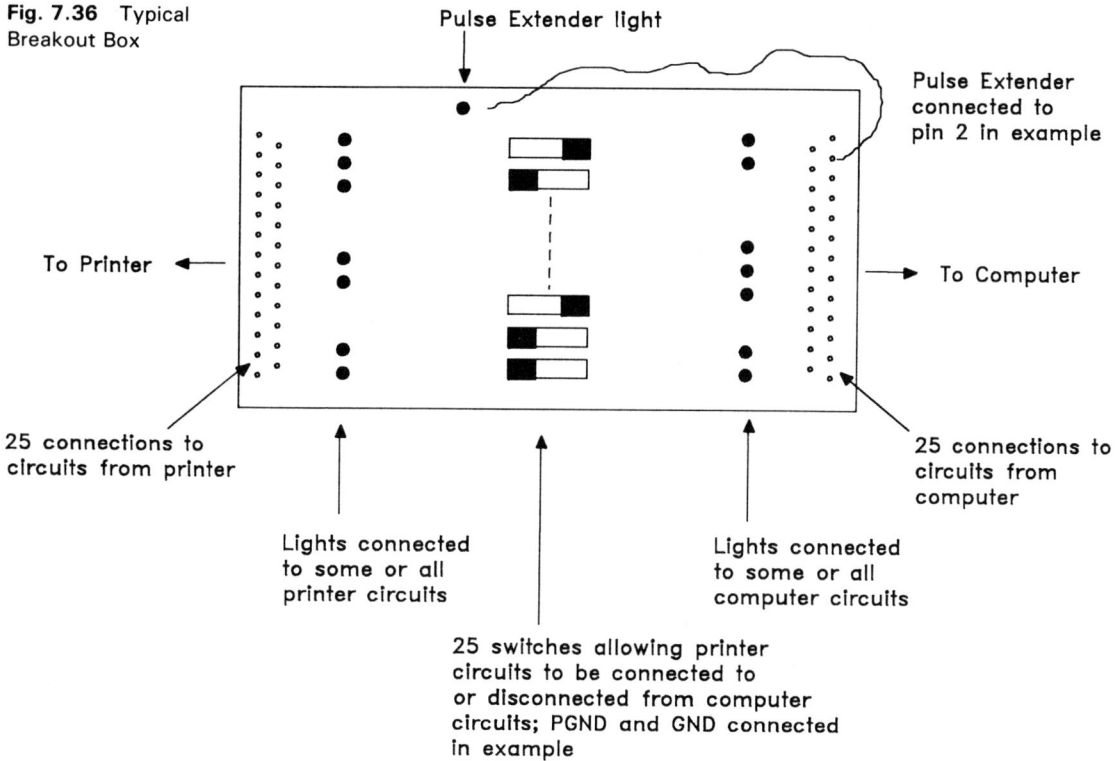

Fig. 7.36 Typical Breakout Box

Pulse Extender light

Pulse Extender connected to pin 2 in example

To Printer

To Computer

25 connections to circuits from printer

25 connections to circuits from computer

Lights connected to some or all printer circuits

Lights connected to some or all computer circuits

25 switches allowing printer circuits to be connected to or disconnected from computer circuits; PGND and GND connected in example

- The printer needs DCD and DSR.
- The computer needs DCD and CTS.

We now need to find the circuit generators implemented by both ends. These might be RTS on pin 4 and DTR on pin 20. Most boxes will have lights permanently wired to these pins on each side, but if not, all boxes should have a test light that can be connected to the four pins, two on each side, in turn.

It would be a fair guess that the printer does not implement RTS because we know that it does not implement CTS. Likewise, it is fair to guess that the computer does implement RTS.

Suppose the tests show the following:

- The printer generates DTR.
- The computer generates RTS.

A possible connection given these facts is shown in Fig. 7.37; this can be tried-out using the box. All switches should be off except for pins 1 and 7, and the "cable" can be constructed by patching between the appropriate pins on either side of the box.

A good trial of the connection is desirable to explore all the possible operating conditions that the real cable will have to work under. There may for example be problems with timing, such that one piece of

**Fig. 7.37** Cable for
printer connection

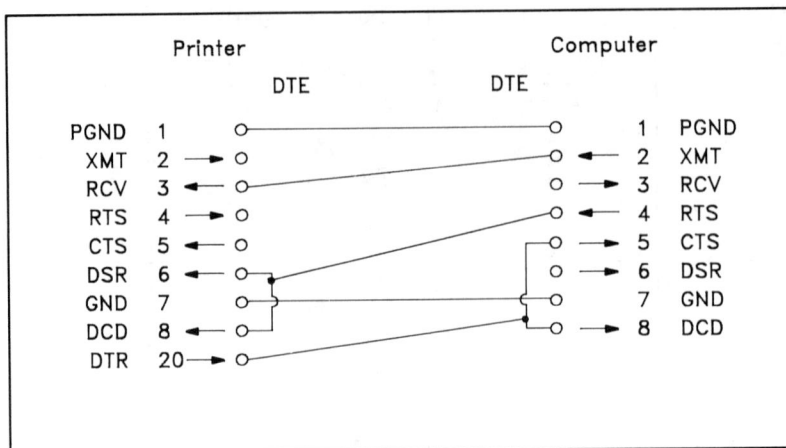

equipment generates a one on a circuit later than the other piece of
equipment needs to see it.

One problem that the test will almost certainly reveal is flow control,
since we have not considered any way in which the printer can indicate
that its buffer is full. Since we have established that the printer does not
implement RTS/CTS flow control – it does not generate RTS nor
require CTS – the most likely alternative is XON/XOFF flow control. In
this method the device sends the ASCII character XOFF to stop the
other device sending, and later sends XON to indicate that it can receive
more data.

The fact that the printer is using this method of flow control could be
established by connecting pin 2 of the printer to the pulse extender, and
observing that the light comes on just before the printer corrupts the
data.

To make the printer work properly we need to connect its pin 2 to pin
3 of the computer, so that the flow control characters can be received by
the computer software. The software may need to be configured to
operate with this different type of flow control, and this may in turn
upset the way in which the computer uses pins 4 and 5. More work with
the box is thus called for.

The Breakout Box can be a very helpful ally to the network manager
in sorting out physical problems. Apart from its "logical" use as shown
above, it may also reveal unexpected things such as devices using spare
pins for flow control, odd clock sources, and timing problems.

## 7.15 Inside the devices

This section is not a course in electronics, but will illustrate in block
form what is inside the DTE and DCE, and what choices may be
implemented.

**Fig. 7.38** DTE internal
components

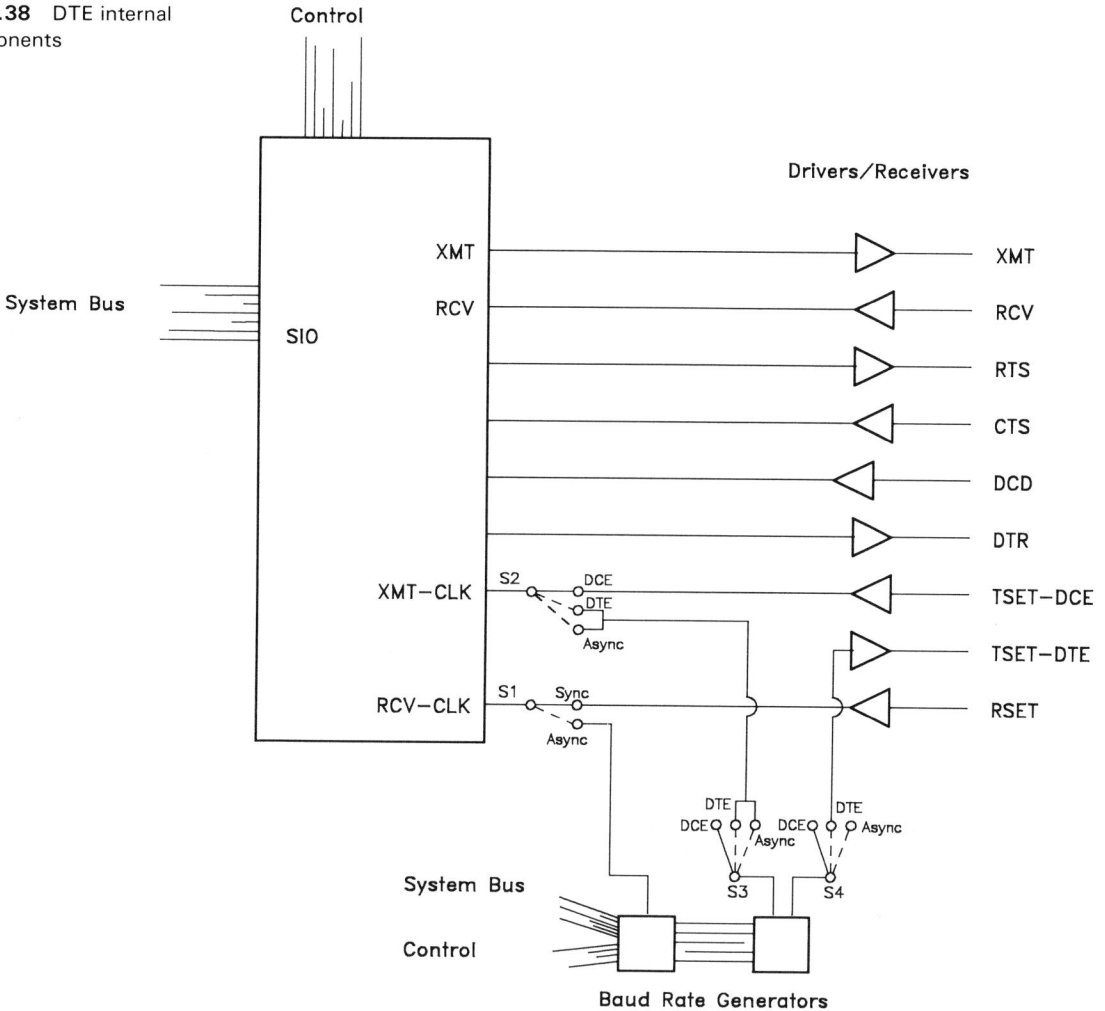

Control

Drivers/Receivers

System Bus

SIO

XMT ▷ XMT

RCV ◁ RCV

RTS ▷ RTS

CTS ◁ CTS

DCD ◁ DCD

DTR ▷ DTR

XMT−CLK — S2 — DCE / DTE / Async ◁ TSET−DCE

▷ TSET−DTE

RCV−CLK — S1 — Sync / Async ◁ RSET

DTE / DCE / Async — S3 / S4 — DTE / DCE / Async

System Bus

Control

Baud Rate Generators

## 7.15.1  DTE

Figure 7.38 shows the inside of a DTE. The major component is the
*Synchronous Input Output* (SIO) chip which is a peripheral device of
the main processor, and which is attached to the processor, and to every
other peripheral, by the system bus. A second connection between the
SIO and the processor allows the SIO to be controlled. Depending on
the instructions presented to the SIO by this control mechanism:

- The SIO can take a character in parallel form from the system bus,
  and present it as a serial character to the outside world.
- The SIO can convert a character received in serial form from the
  outside world into parallel form, and place it on the system bus.

Despite its name the SIO can usually be set to generate and receive
asynchronous as well as synchronous characters.

The SIO is the interface between the internal mechanisms of the equipment and the external mechanisms. It can be thought of as converting the processor style to the V.24 style; and implements many of the V.24 circuits. The exact number will depend on the type of SIO used. Those circuits that it does not implement need to be driven from other components connected to the system bus and control mechanisms.

Between the SIO and the ISO 2110 connector is a set of driver/receiver components, one for each circuit depending on whether it is a generator or a load. The driver/receiver converts from the electrical characteristics of the SIO to the electrical characteristics required at the connector. In a typical DTE they would generate or receive V.28 electrical characteristics. For most of the circuits this is all that is required. However, the clock circuits present a little more difficulty.

The SIO always needs to be provided with clock signals, whether these are provided by the external equipment or internally. When clock signals must be generated within the equipment, then this is done by *Baud Rate Generators* (BRGs). The Baud Rate Generators provide the equal on and off signal required, and can be programmed by the system bus to do this at the required baud rate.

Taking the receive clock first of all, this is provided by the external equipment for synchronous data, and the Baud Rate Generator is not needed. For asynchronous data the SIO must be provided with a clock, so it is connected to the Baud Rate Generator. This choice is made with switch S1.

For the transmit clock, for synchronous data the clock is usually supplied by the external equipment (TSET-DCE). The clock may alternatively be provided by this equipment (TSET-DTE) in which case the Baud Rate Generator must be connected both to the external equipment and to the SIO. Two outputs are taken from the Baud Rate Generator to do this because the SIO must start to generate the bits on circuit 103 half a bit time before the TSET-DCE circuit indicates that the DCE may sample them. For asynchronous data the SIO must again be connected to the Baud Rate Generator. These choices are made using switches S2, S3, S4 and they should all be adjusted to the same selection.

It is important that clock signals are not carried to the connector unless it is necessary. For instance, with asynchronous data the clock should only be carried to the SIO, not the connector as well. This is to prevent the possible induction of the signal into adjacent circuits at the connector. It also prevents the almost inevitable induction of signals if the cabling practice is poor.

For example, if switch S4 was set to "sync DTE" and the data was asynchronous, then an unnecessary clock would be carried to pin 24 of the connector. This in itself may be enough to induce false signals into the other circuits. If the DTE-to-DCE connection was made with a ribbon cable, and especially if pin 24 was not connected in the DCE, then data corruption would be extremely likely.

### 7.15.2 Programmable logic

Much of the work of configuring the hardware can be handled by software, and as shown in the previous section, many types of SIOs can be programmed through the system bus. Parameters such as mode (synchronous or asynchronous), parity, and asynchronous framing characteristics (number of stop bits for example), can all be selected in this way. These choices do not have to be made by switches, so configuration is faster and easier. The clock select switches can also be made software-selectable by the use of programmable registers. These are effectively switches that are driven from the system bus and by using them the entire configuration of the unit can be done from an operator's screen.

SIOs are available that include the Baud Rate Generators on the same chip, and it seems likely that the entire input/output circuitry will eventually be reduced to a single component that has no manual selections.

### 7.15.3 DCE

The SIO chip has no particular orientation with regard to DCE or DTE, it just has input and output circuits which are driven from the system bus. To make a DCE it is basically only necessary to change the labels on Fig. 7.38 so that XMT and RCV are swapped, DTR and DCD are swapped, and so on. The clocks are a little more difficult since RSET is now an output, and the chip must be clocked half a bit time before the DTE can sample circuit 104.

This allows for the possibility of designing an interface that is configurable for either DTE or DCE. The most difficult part of such a design is deciding which clocks are required when, and this configuration is probably best implemented in programmable logic. Two connectors are also required, male for DTE and female for DCE, and the unused one is probably best "turned off" through logic to reduce stray induction.

## 7.16   RS-232-C

This is a standard defined by the Electrical Industries Association (EIA) of America. It is similar in scope to the combination of V.24, V.28 and ISO 2110 and shares most of the details of the CCITT recommendations. RS-232-C does not define a connector, though it requires a 25-way male and female arrangement as for ISO 2110, and refers to the familiar D-type mechanical design.

As with V.28, there is no statement in RS-232-C of the possible distance that the cable can span between the DCE and DTE. The

| Pin | RS–232–C | V.24 |
|---|---|---|
| 1 | AA | Protective Ground |
| 2 | BA | Transmitted Data |
| 3 | BB | Received Data |
| 4 | CA | Request to Send |
| 5 | CB | Clear to Send |
| 6 | CC | Data Set Ready |
| 7 | AB | Signal Ground |
| 8 | CF | Received Line Signal Detector |
| 9 | | |
| 10 | | |
| 11 | | |
| 12 | SCF | Secondary Received Line Signal Det. |
| 13 | SCB | Secondary Clear to Send |
| 14 | SBA | Secondary Transmitted Data |
| 15 | DB | Trans. Signal Element Timing – DCE source |
| 16 | SBB | Secondary Received Data |
| 17 | DD | Receive Signal Element Timing |
| 18 | | |
| 19 | SCA | Secondary Request to Send |
| 20 | CD | Data Terminal Ready |
| 21 | CG | Signal Quality Detector |
| 22 | CE | Ring Indicator |
| 23 | CH | Data Signal Rate Selector |
| 24 | DA | Trans. Signal Element Timing – DTE source |
| 25 | | |

**Fig. 7.39** RS-232-C circuits

standard does mention cables of fifty feet in length, and this is commonly taken to be a safe length. As explained in the section on V.28, this is very variable depending on cable quality.

The circuits in RS-232-C are named with two or three alphabetic characters. Figure 7.39 shows the pin assignment of these circuits, and the corresponding V.24 names.

RS-232-C does not define circuits for loopback testing, so pin 21 is assigned to another use.

## 7.17 X.21

This recommendation describes a set of interchange circuits and the procedures governing the use of the circuits. It refers to ISO 4903 for the connector and recommendations X.26 and X.27 for the electrical characteristics. Figure 7.40 shows the circuits and their single-character abbreviations.

X.21 uses a state transition system for controlling the DTE-to-DCE conversation; thus the meaning of any particular combination of signals depends on what has happened previously. One of the states is "Data Transfer" which is signified by both Control and Indication being on,

Fig. 7.40   X.21 circuits

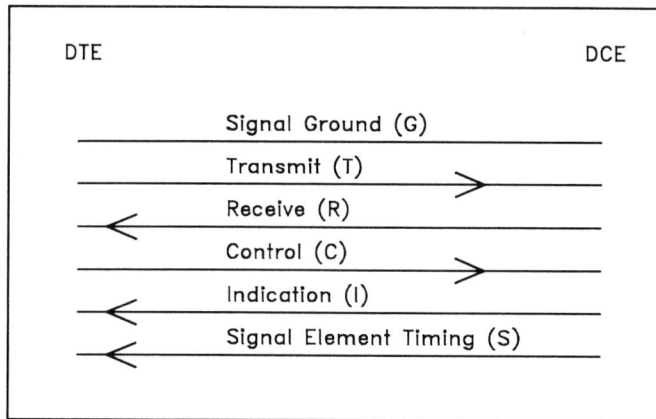

and in this state synchronous data is passed between the DTE and DCE using the Transmit and Receive circuits. Both data streams are clocked by the same clock circuit, Signal Element Timing, which is located in the DCE.

A change in state is indicated by the Control and Indication circuits changing, and the state of the Transmit and Receive circuits is significant in determining the new state to enter. Thus the Transmit and Receive circuits are used to control the interface as well as to pass data. A large number of states (more than twenty) are defined for the interface, including ones for loopback testing of X.21 modems.

There is an optional extra circuit for byte timing, which is simply an extra clock that ticks on every eighth tick of the normal clock. This is used by some DTEs in the controlling states.

The connector defined in ISO 4903 is a fifteen-way D-type similar in appearance to the more common twenty-five-way type. The usual arrangement of male for DTE and female for DCE applies.

Two electrical recommendations are used with X.21: X.26 at speeds up to 9600 bps; and X.27 for higher speeds. X.26 (which is the same as V.10) is similar to V.28, and uses the Signal Ground circuit as a reference potential. It is based on a five volt electrical system.

X.27 (which is the same as V.11) has two conductors for each circuit. Thus for example, there is Transmit A (TA) and Transmit B (TB). These conductors are twisted together in the cable, and form a reference potential for each other. If A is positive with respect to B then one level is being generated, and if B is positive with respect to A then the other level is being generated. The Signal Ground circuit is not required. This type of electrical system is much less prone to corruption from external − and indeed internal − sources than V.28, and can be used at speeds up to 64K bps without problem.

X.21 defines states for taking the connection from idle, through call establishment to data transfer, and back down again. Where a higher-level protocol such as X.25 is used, these states are not all necessary and many installations only implement a subset of them.

The use of X.21 is very limited despite being referred to in X.25, and there are only a small number of modems and other items of equipment that use it. A second recommendation, X.21 bis, defines an X.21 type of service but using the normal V.24, V.28, and ISO 2110 interface. This allows existing modems and wiring to be used, but gives the benefits of the X.21 approach.

Recommendations X.20 and X.20 bis are similar to X.21 and X.21 bis but for asynchronous data.

# APPENDIX A

## Physical Interchange Circuits

| ISO 2110 pin number | RS-232-C circuit | V.24 Circuit | | | Direction DTE DCE | |
|---|---|---|---|---|---|---|
| 1 | AA | | Protective Ground | (PGND) | | 1 |
| 2 | BA | 103 | Transmitted Data | (XMT) | → | 2 |
| 3 | BB | 104 | Received Data | (RCV) | ← | 3 |
| 4 | CA | 105 | Request To Send | (RTS) | → | 4 |
| 5 | CB | 106 | Clear To Send | (CTS) | ← | 5 |
| 6 | CC | 107 | Data Set Ready | (DSR) | ← | 6 |
| 7 | AB | 102 | Signal Ground | (GND) | | 7 |
| 8 | CF | 109 | Data Carrier Detect | (DCD) | ← | 8 |
| 9 | | | | | | 9 |
| 10 | | | | | | 10 |
| 11 | | | | | | 11 |
| 12 | SCF | 122 | Secondary Data Carrier Detect | (SDCD) | ← | 12 |
| 13 | SCB | 121 | Secondary Clear To Send | (SCTS) | ← | 13 |
| 14 | SBA | 118 | Secondary Transmitted Data | (SXMT) | → | 14 |
| 15 | DB | 114 | Transmit Signal Element Timing | (TSET-DCE) | ← | 15 |
| 16 | SBB | 119 | Secondary Received Data | (SRCV) | ← | 16 |
| 17 | DD | 115 | Receive Signal Element Timing | (RSET) | ← | 17 |
| 18 | | 141 | Local Loopback | (L3) | → | 18 |
| 19 | SCA | 120 | Secondary Request To Send | (SRTS) | → | 19 |
| 20 | CD | 108 | Data Terminal Ready | (DTR) | → | 20 |
| 21 | * | 140 | Remote Loopback | (L2) | → | 21 |
| 22 | CE | 125 | Ring Indicator | (RI) | ← | 22 |
| 23 | CH | 111 | Data Signalling Rate Selector | (SEL) | → | 23 |
| 24 | DA | 113 | Transmit Signal Element Timing | (TSET-DTE) | → | 24 |
| 25 | | 142 | Test Indicator | (TI) | ← | 25 |

* RS-232-C does not define circuits for loopback testing.
Pin 21 is used for Signal Quality Detector (CG).

# APPENDIX B

## X.25 (1980) Frame and Packet Formats

### B 1.  Frame Formats

### B 1.1  Data on the Circuit

7E 7E 7E 7E 7E     Frame    7E Frame 7E 7E    Frame    7E 7E

- At least one flag byte between frames
- Line always busy sending flags or frames
- Flag is hex 7E - binary 01111110
- Frames are of different lengths
- Frames are bit-stuffed to avoid occurrence of six binary ones

### B 1.2  General Frame Layout

7E 7E 7E 7E 7E          Frame           7E 7E

| Either | Address | Control | | FCS |
|--------|---------|---------|---|-----|
| Or | Address | Control | Information | FCS |

- Address is eight bits
- Address is 3 for DCE → DTE commands or DTE → DCE responses
- Address is 1 for DTE → DCE commands or DCE → DTE responses
- Control is eight bits and shows frame type (see below)
- Information field only present in I-frames and CMDR/FRMR frames
- Information field carries the data to be transmitted and may be any length up to an agreed maximum
- FCS is sixteen bits

## B 1.3 Coding of Control Field

```
Contents of field
high order      low order
8 7 6 5 4 3 2 1
```

| | | | |
|---|---|---|---|
| 0 | I | Information frame | |
| 0 1 | S | Supervisory frame | |

| | | |
|---|---|---|
| 0 0 0 1 | RR | Receiver Ready |
| 0 1 0 1 | RNR | Receiver Not Ready |
| 1 0 0 1 | REJ | Reject |

| | | |
|---|---|---|
| 1 1 | U | Unnumbered frame |

| | | | |
|---|---|---|---|
| 0 0 0 | 1 1 1 1 | SARM | Set Asynchronous Response Mode |
| 0 0 1 | 1 1 1 1 | SABM | Set Asynchronous Balanced Mode |
| 0 1 0 | 0 0 1 1 | DISC | Disconnect |
| 0 1 1 | 0 0 1 1 | UA | Unnumbered Acknowledgement |
| 1 0 0 | 0 1 1 1 | CMDR/FRMR | Command/Frame Reject |

- CMDR/FRMR choice is determined by whether Response mode or Balanced mode is in effect
- The coding for the SARM command is the same as for the DM response

## B 1.4 Information (I) Frame

The contents of the Control field for the I-frame are as follows:

```
Contents of field
high order      low order
8 7 6 5 4 3 2 1
R R R P S S S 0
```

Where -  
    RRR  represents the $N(R)$ value  
    P  represents the Poll bit  
    SSS  represents the $N(S)$ value

The frame portion between the Control field and the FCS contains the layer three packet information.

## B 1.5    Receiver Ready (RR) Frame

The contents of the Control field for the RR frame are as follows:

```
Contents of field
high order      low order
8 7 6 5 4 3 2 1
R R R P 0 0 0 1
```

Where -      RRR     represents the N(R) value
             P       represents the Poll/Final bit

## B 1.6    Receiver Not Ready (RNR) Frame

The contents of the Control field for the RNR frame are as follows:

```
Contents of field
high order      low order
8 7 6 5 4 3 2 1
R R R P 0 1 0 1
```

Where -      RRR     represents the N(R) value
             P       represents the Poll/Final bit

## B 1.7    Reject (REJ) Frame

The contents of the Control field for the REJ frame are as follows:

```
Contents of field
high order      low order
8 7 6 5 4 3 2 1
R R R P 1 0 0 1
```

Where -      RRR     represents the N(R) value
             P       represents the Poll/Final bit

## B 1.8    Set Asynchronous Response Mode (SARM) Frame

The contents of the Control field for the SARM or DM frame are as follows:

```
Contents of field
high order    low order
8 7 6 5 4 3 2 1
0 0 0 P 1 1 1 1
```

Where -  P represents the Poll/Final bit

## B 1.9    Set Asynchronous Balanced Mode (SABM) Frame

The contents of the Control field for the SABM frame are as follows:

```
Contents of field
high order    low order
8 7 6 5 4 3 2 1
0 0 1 P 1 1 1 1
```

Where -  P represents the Poll bit

## B 1.10   Disconnect (DISC) Frame

The contents of the Control field for the DISC frame are as follows:

```
Contents of field
high order    low order
8 7 6 5 4 3 2 1
0 1 0 P 0 0 1 1
```

Where -  P represents the Poll bit

## B 1.11   Unnumbered Acknowledgement (UA) Frame

The contents of the Control field for the UA frame are as follows:

```
Contents of field
high order    low order
8 7 6 5 4 3 2 1
0 1 1 F 0 0 1 1
```

Where -  F represents the Final bit

## B 1.12   Command Error / Frame Reject (CMDR/FRMR) Frame

The contents of the Control field for the CMDR/FRMR frame are as follows:

```
Contents of field
high order     low order
8 7 6 5 4 3 2 1
1 0 0 F 0 1 1 1
```

Where -  F represents the Final bit

The contents of the Information field are as follows:

```
Contents of 1st byte     Contents of 2nd          Contents of 3rd
high          low        high          low         high          low
8 7 6 5 4 3 2 1          8 7 6 5 4 3 2 1           8 7 6 5 4 3 2 1
C C C C C C C C          R R R N S S S 0           0 0 0 0 D D D D
```

Where -  C...C    Represents the Control field of the rejected frame

RRR    Represents the current receive state variable in the component reporting the error

N    Is 0 if the frame is CMDR.  For an FRMR N is 1 if the rejected frame is a response, and N is 0 if the rejected frame is a command.

SSS    Represents the current send state variable in the component reporting the error.

DDDD    Is 0000 if the frame is a CMDR.  If the frame is an FRMR then the bits indicate the reason for the rejection as follows.  In all cases the Control field of the rejected frame is returned in the first byte.

```
                   4 3 2 1
```

|  |  |
|---|---|
| 1 | The Control field is not valid. |
| 1 1 | The combination of the Control field and Information field is not valid. |
| 1 | The Information field exceeds the agreed maximum size. |
| 1 | The N(R) of the rejected frame is not valid. |

## B 2. Packet Formats

### B 2.1 Basic Packet Layout

The packet is wholly contained in the Information field of an Information Frame. The field is subdivided into the fields of the packet. Note that the Control field of the frame has a low order bit value of zero to indicate the Information Frame, and therefore the presence of a packet.

Packet layout:

either           GFI  LCG  LCN  TYPE

or               GFI  LCG  LCN  TYPE  DATA

- The DATA field is present in most types of packet and carries data relevent to the packet.
- The GFI and LCG form the first byte of the packet
- The LCN forms the second byte
- The TYPE forms the third

### B 2.1.1 GFI format

The GFI (Group Field Identifier) is a four bit field arranged as follows:

```
High order        Low order
          4  3  2  1

          Q  D  M  M
```

where -  Q   Represents the Qualifier bit (Q-bit) - it is set to 0 in packets of types other than Data
         D   Represents the Delivery Confirmation bit (D-bit) - it is set to 0 in packets of types other than Data or Call Request / Incoming Call
         MM  Represents the modulus used for packet sequnce numbering - 01 for modulus 8 (0-7) and 10 for modulus 128 (0-127). 11 is also defined for future extensions to the available packet formats.

### B 2.1.2 LCG format

The LCG (Logical Channel Group) is a four bit field coded in binary that contains the first hexadecimal digit of the Logical Channel Identifier. Along with the LCN, this field defines which of the possible calls on the link this packet is associated with.

## B 2.1.3 LCN Format

The LCN (Logical Channel Number) is an eight bit field coded in binary that contains the second and third hexadecimal digits of the Logical Channel Identifier. Along with the LCG, this field defines which of the possible calls on the link this packet is associated with.

## B 2.1.4 TYPE Format

The TYPE field indicates what type of packet this is, and therefore indicates what fields - if any - follow it. It is encoded as follows. The packets are shown with both DTE-to-DCE and DCE-to-DTE naming.

| Field contents high order    low order 8 7 6 5 4 3 2 1 | DCE-to-DTE | DTE-to-DCE |
|---|---|---|
|              0 | Data | Data |
|       0 0 0 0 1 | RR | RR   \| |
|       0 0 1 0 1 | RNR | RNR  \| modulo 8 |
|       0 1 0 0 1 | | REJ   \| |
| 0 0 0 0 0 0 0 1 | RR | RR   \| |
| 0 0 0 0 0 1 0 1 | RNR | RNR  \| modulo 128 |
| 0 0 0 0 1 0 0 1 | | REJ   \| |
| 0 0 0 1 1 0 1 1 | Reset indication | Reset request |
| 0 0 0 1 1 1 1 1 | Reset confirmation | Reset confirmation |
| 0 0 0 0 1 0 1 1 | Incoming call | Call request |
| 0 0 0 0 1 1 1 1 | Call connected | Call accepted |
| 0 0 0 1 0 0 1 1 | Clear indication | Clear request |
| 0 0 0 1 0 1 1 1 | Clear confirmation | Clear confirmation |
| 1 1 1 1 1 0 1 1 | Restart indication | Restart request |
| 1 1 1 1 1 1 1 1 | Restart confirmation | Restart confirmation |
| 0 0 1 0 0 0 1 1 | Interrupt | Interrupt |
| 0 0 1 0 0 1 1 1 | Interrupt confirmation | Interrupt confirmation |

## B 2.2   Data Packet

Packet layout for modulo 8 working:

```
1st byte    2nd byte   -3rd byte--     4th-Nth byte
GFI  LCG     LCN       P(R) M P(S)        Data
```

The third byte is encoded as follows:

```
High order     low order bit
8 7 6 5 4 3 2 1

R R R M S S S 0
```

where -    RRR   represents the P(R) value
           M     represents the More data bit
           SSS   represents the P(S) value

Subsequent bytes - up to the agreed  maximum - contain the application data.

For modulo 128 working the packet layout is as follows:

```
1st byte    2nd byte   3rd byte   4th byte   5th-Nth byte

GFI LCG      LCN        P(S)       P(R) M      Data
```

The third byte is encoded as follows:

```
High order     low order bit
8 7 6 5 4 3 2 1

S S S S S S S 0
```

where -  S...S represents the P(S) value.

The fourth byte is encoded as follows:

```
High order     low order bit
8 7 6 5 4 3 2 1

R R R R R R R M
```

where -    R...R      represents the P(R) value
           M          represents the More data bit

Subsequent bytes - up to the agreed maximum - contain the application data.

## B 2.3  RR (Receiver Ready) Packet

Packet layout for modulo 8 working:

```
1st byte    2nd byte    3rd byte-

GFI LCG       LCN      P(R) TYPE
```

The layout of the third byte is as follows:

```
High order    low order bit
8 7 6 5 4 3 2 1

R R R 0 0 0 0 1
```

where -  RRR represents the P(R) value

For modulo 128 working the packet layout is as follows:

Packet layout:

```
1st byte    2nd byte    3rd byte    4th byte

GFI  LCG      LCN        TYPE        P(R)
```

•  The packet TYPE is 00000001

The layout of the fourth byte is as follows:

```
High order    low order bit
8 7 6 5 4 3 2 1

R R R R R R R 0
```

where -  RRRRRRR represents the P(R) value.

## B 2.4  RNR (Receiver Not Ready) Packet

Packet layout for modulo 8 working:

```
1st byte    2nd byte    3rd byte-

GFI LCG       LCN      P(R) TYPE
```

The layout of the third byte is as follows:

```
High order    low order bit
8 7 6 5 4 3 2 1

R R R 0 0 1 0 1
```

where -  RRR represents the P(R) value

For modulo 128 working the packet layout is as follows:

Packet layout:

| 1st byte | 2nd byte | 3rd byte | 4th byte |
|----------|----------|----------|----------|
| GFI LCG  | LCN      | TYPE     | P(R)     |

- The packet TYPE is 00000101

The layout of the fourth byte is as follows:

```
High order      low order bit
8 7 6 5 4 3 2 1

R R R R R R R 0
```

where - RRRRRRR represents the P(R) value.

## B 2.5  REJ  (Reject) Packet

Packet layout for modulo 8 working:

| 1st byte | 2nd byte | 3rd byte- |
|----------|----------|-----------|
| GFI LCG  | LCN      | P(R) TYPE |

The third byte is encoded as follows:

```
High order      low order bit
8 7 6 5 4 3 2 1

R R R 0 1 0 0 1
```

where - RRR represents the P(R) value.

For modulo 128 working the packet layout is as follows:

| 1st byte | 2nd byte | 3rd byte | 4th byte |
|----------|----------|----------|----------|
| GFI LCG  | LCN      | TYPE     | P(R)     |

- The packet TYPE is 00001001

The fourth byte is encoded as follows:

```
High order      low order bit
8 7 6 5 4 3 2 1

R R R R R R R 0
```

where - RRRRRRR represents the P(R) value.

221

## B 2.6  Reset request / Reset Indication Packet

Packet layout:

```
1st byte   2nd byte   3rd byte   --4th byte-   ---5th byte----

GFI LCG    LCN        TYPE       Reset cause   Diagnostic code
```

- The packet TYPE is 00011011
- If the Diagnostic code is not required the field may be left off.
- Cause and Diagnostic codes are shown in Section 2.17 of this Appendix.


## B 2.7  Reset Confirmation Packet

Packet layout:

```
1st byte   2nd byte   3rd byte

GFI LCG      LCN        TYPE
```

- The packet TYPE is 00011111.


## B 2.8  Call Connected / Call Accepted Packet

Packet layout:

```
1st byte 2nd byte 3rd byte   --------4th byte-------------

GFI LCG    LCN       TYPE     Calling DTE    Called DTE
                             address length  address length

5th byte--Nth byte    N+1st------Mth byte   ---M+1st byte--

Called DTE address    Calling DTE address   Facility length

M+2nd---Xth byte

   Facilities
```

- The packet TYPE is 00001111
- If neither of the addresses nor the Facilities fields are required then the 4th to the Xth byte may be left off of the packet.
- Either or both addresses may be absent in which case the address length is zero and the address field is not inserted.
- Addresses are encoded four bits per digit.
- If the sum of the address lengths is an odd number, such that the Facility length field would not be aligned on a byte, then the final address is padded with four 0 bits.

- The Facilities field may be absent in which case the length is zero and the field is not inserted.
- The maximum length of the Facilities field is 63 bytes, hence the two high order bits of the Facility length field are always 0.
- The coding of the Facilities field is shown in Section 2.16 of this Appendix
- If the Fast Select facility is used, then the two Address length fields and the Facilities length field are required. The X+1st byte to the Yth byte may then contain Called User Data up to a maximum length of 128 bytes. The length fields are required whether or not Called User Data is present.

## B 2.9  Incoming Call / Call Request Packet

Packet layout:

```
1st byte 2nd byte 3rd byte  -----------4th byte-----------

GFI LCG      LCN      TYPE      Calling DTE    Called DTE
                               address length  address length

5th byte---Nth byte  N+1st------Mth byte   ---M+1st byte--

Called DTE address   Calling DTE address   Facility length

M+2nd---Xth byte    X+1st---Yth byte

     Facilities      Call User data
```

- The packet TYPE is 00001011
- Either or both addresses may be absent in which case the address length is zero and the address field is not inserted.
- Addresses are encoded four bits per digit.
- If the sum of the address lengths is an odd number, such that the Facility length field would not be aligned on a byte, then the final address is padded with four 0 bits.
- The Facilities field may be absent in which case the length is zero and the field is not inserted.
- The maximum length of the Facilities field is 63 bytes, hence the two high order bits of the Facility length field are always 0.
- The coding of the Facilities field is shown in Section 2.16 of this Appendix
- The Call User Data field may be absent in which case the field is not inserted.
- The Call User Data field has a maximum length of 16 bytes or 128 bytes if the Fast Select facility is used.

## B 2.10 Clear Indication / Clear Request Packet

Packet layout for calls not using the Fast Select facility:

```
1st byte 2nd byte 3rd byte   ---4th byte---  ----5th byte---

GFI LCG    LCN       TYPE     Clearing cause  Diagnostic code
```

- The packet TYPE is 00010011
- The contents of the Clearing cause field and the Diagnostic code field are shown in Section 2.17 of this Appendix.
- If the Diagnostic code field is not required then it may be left off the packet

When the Fast Select facility is used the packet layout is as follows:

```
1st byte    2nd byte    3rd byte    ---4th byte---

GFI LCG       LCN         TYPE       Clearing cause

----5th byte---  ----------6th byte-----------

Diagnostic code  Calling DTE     Called DTE
                 address length  address length

7th byte--Nth byte    N+1st------Mth byte    ---M+1st byte--

Called DTE address    Calling DTE address    Facility length

M+2nd---Xth byte    X+1st---Yth byte

   Facilities        Call User data
```

- The TYPE is 00010011
- The Clearing causes and Diagnostic codes are shown in Section 2.17 of this Appendix.
- The presence of addresses is not yet defined so the two address length fields must both be filled with zeros, and the address fields are not present.
- The coding of the Facilities field is shown in Section 2.16 of this Appendix
- The Clear User Data field is optional and has a maximum length of 128 bytes.

## B 2.11   Clear Confirmation Packet

Packet layout:

```
1st byte    2nd byte   3rd byte

GFI LCG       LCN        TYPE
```

- The packet TYPE is 00010111

## B 2.12   Restart Indication / Restart Request Packet

Packet layout:

```
1st byte   2nd byte   3rd byte   ---4th byte--   ----5th byte---

GFI LCG      LCN        TYPE     Restart Cause   Diagnostic code
```

- The LCG and LCN fields are set to all zeros to indicate a Restart of the procedures rather than an action on a particular Virtual Circuit.
- The packet TYPE is 11111011
- The Restart Cause and Diagnostic codes are shown in Section 2.17 of this Appendix.

## B 2.13   Restart Confirmation Packet

Packet layout:

```
1st byte    2nd byte   3rd byte

GFI LCG       LCN        TYPE
```

- The LCG and LCN fields are set to all zeros to indicate a Restart of the procedures rather than an action on a particular Virtual Circuit.
- The packet TYPE is 11111111.

## B. 2.14   Interrupt Packet

Packet layout:

```
1st byte    2nd byte   3rd byte   -----4th byte------

GFI LCG       LCN        TYPE      Interrupt User Data
```

- The packet TYPE is 00100011

## B 2.15   Interrupt Confirmation Packet

Packet layout:

```
1st byte   2nd byte   3rd byte

GFI LCG      LCN        TYPE
```

  • The packet TYPE is 00100111

## B 2.16   Coding of Facilities field

The Facility field is used in Call Connected / Call Accepted, Incoming Call / Call Request, and Clear Indication / Clear Request packets.  It indicates features that are required for the call.

The field is divided into a number of facility requests, each consisting of a single byte facility identifier followed by a number of facility parameters.

For example, if three features are requested, with one, three, and two bytes of parameters respectively, then the Facility field would appear as: FPFPPPFPP.

### B 2.16.1   Facility Classes

Each facility belongs to a class, and the class indicates how many bytes make up the facility request.  The class is encoded in the two high order bits of the facility identifier as follows:

```
                                 Facility identifier byte
                                 high order    low order
```

Class A -   facilities with one         `0  0  I  I  I  I  I  I`
            parameter byte

Class B -   facilities with two         `0  1  I  I  I  I  I  I`
            parameter bytes

Class C -   facilities with three       `1  0  I  I  I  I  I  I`
            parameter bytes

Class D -   facilities with more than   `1  1  I  I  I  I  I  I`
            three parameter bytes

where - `I...I` represents the facility identifier

In the case of Class D facilities the facility identifier is followed by a byte containing the number of parameter bytes that follow.

## B 2.16.2   Facility marker

This special facility, often called the National Options Marker, marks the end of X.25 facilities and the start of special non-X.25 facilities that may be offered by the network administration.

Class A    first byte      0 0 0 0 0 0 0 0
           second byte     0 0 0 0 0 0 0 0      if a facility on the
                                                source network

           second byte     1 1 1 1 1 1 1 1      if a facility on the
                                                destination network

## B 2.16.3   Reverse Charging and Fast Select

Class A    first byte      0 0 0 0 0 0 0 1
           second byte     F F 0 0 0 0 0 R

where - R          is 0 for Reverse charging not selected
                   is 1 for Reverse charging selected

        FF         is 00 or 01 for Fast Select not requested
                   is 10 for Fast Select with no restriction on response
                   is 11 for Fast Select with restriction

## B 2.16.4   Closed User Group

Class A    first byte      0 0 0 0 0 0 1 1
           second byte     M M M M N N N N

where -  MMMM       is the first digit of the selected group
         NNNN       is the second digit

For a Bilateral Closed User Group:

Class B    first byte      0 1 0 0 0 0 0 1
           second byte     M M M M N N N N
           third byte      X X X X Y Y Y Y

where -  MMMM       is the first digit of the selected group
         NNNN       is the second digit
         XXXX       is the third digit
         YYYY       is the fourth digit

## B 2.16.5　RPOA selection

| Class B | first byte | 0 1 0 0 0 1 0 0 |
|---------|------------|-----------------|
|         | second byte | M M M M N N N N |
|         | third byte | X X X X Y Y Y Y |

where -　MMMM　is the first digit of the DNIC
　　　　NNNN　is the second digit
　　　　XXXX　is the third digit
　　　　YYYY　is the fourth digit

## B 2.16.6　Window size

| Class B | first byte | 0 1 0 0 0 0 1 1 |
|---------|------------|-----------------|
|         | second byte | 0 T T T T T T T |
|         | third byte | 0 C C C C C C C |

where -　T...T　is the window size in the direction from the called DTE
　　　　C...C　is the window size in the direction to the called DTE

The values are encoded in binary.

## B 2.16.7　Packet size

| Class B | first byte | 0 1 0 0 0 0 1 0 |
|---------|------------|-----------------|
|         | second byte | 0 0 0 0 T T T T |
|         | third byte | 0 0 0 0 C C C C |

where -　TTTT　is the packet size in the direction from the called DTE
　　　　CCCC　is the packet size in the direction to the called DTE

both are encoded -　0100　for packet size　　　16
　　　　　　　　　　0101　for packet size　　　32
　　　　　　　　　　0110　for packet size　　　64
　　　　　　　　　　0111　for packet size　　128
　　　　　　　　　　1000　for packet size　　256
　　　　　　　　　　1001　for packet size　　512
　　　　　　　　　　1010　for packet size　1024

## B 2.16.8    Throughput class

| Class A | first byte | 0 0 0 0 0 0 1 0 |
|---------|------------|-----------------|
|         | second byte | T T T T C C C C |

where -  TTTT    is the class in the direction from the called DTE

CCCC    is the class in the direction to the called DTE

| both are encoded - | 0011 | for | 75 bit/s |
|--------------------|------|-----|----------|
|                    | 0100 | for | 150 |
|                    | 0101 | for | 300 |
|                    | 0110 | for | 600 |
|                    | 0111 | for | 1200 |
|                    | 1000 | for | 2400 |
|                    | 1001 | for | 4800 |
|                    | 1010 | for | 9600 |
|                    | 1011 | for | 19200 |
|                    | 1100 | for | 48000 |

## B 2.17    Cause and Diagnostic codes

Cause and Diagnostic codes are used in Reset Request / Reset Indication, Clear Request / Clear Indication, and Restart Request / Restart Indication packets.

## B 2.17.1    Reset Cause and Diagnostics

Resetting Cause byte
high order          low order

| | |
|---|---|
| 0 0 0 0 0 0 0 0 | Reset request originated from DTE |
| 0 0 0 0 0 0 0 1 | Out of Order (PVC only) |
| 0 0 0 0 0 0 1 1 | Remote Procedure Error |
| 0 0 0 0 0 1 0 1 | Local Procedure Error |
| 0 0 0 0 0 1 1 1 | Network Congestion |
| 0 0 0 0 1 0 0 1 | Remote DTE operational (PVC only) |
| 0 0 0 0 1 1 1 1 | Network operational (PVC only) |
| 0 0 0 1 0 0 0 1 | Destination is not compatible with this call |

When the Resetting Cause field is zero one of the DTEs has generated a Reset request. In this case the Diagnostic Code is generated by the DTE and is passed unchanged to the remote DTE. Other Resetting Causes are generated by the network, and the standard Diagnostic Codes shown in section 2.17.4 of this appendix are used.

## B 2.17.2 Clear Cause and Diagnostics

Clearing Cause byte
high order      low order

| | |
|---|---|
| 0 0 0 0 0 0 0 0 | Clear request originated from DTE |
| 0 0 0 0 0 0 0 1 | Number busy |
| 0 0 0 0 0 0 1 1 | Facility request is invalid |
| 0 0 0 0 0 1 0 1 | Network congestion |
| 0 0 0 0 1 0 0 1 | Out of Order |
| 0 0 0 0 1 0 1 1 | Access is barred |
| 0 0 0 0 1 1 0 1 | DTE not obtainable |
| 0 0 0 1 0 0 0 1 | Remote Procedure Error |
| 0 0 0 1 0 0 1 1 | Local Procedure Error |
| 0 0 0 1 0 1 0 1 | RPOA Out of Order |
| 0 0 0 1 1 0 0 1 | Reverse Charging acceptance has not been agreed with network administration |
| 0 0 1 0 0 0 0 1 | Destination is not compatible with this call |
| 0 0 1 0 1 0 0 1 | Fast Select acceptance has not been agreed with the network administration |

When the Clearing Cause is zero one of the DTEs has generated a Clear request. In this case the Diagnostic Code is generated by the DTE and is passed unchanged to the remote DTE. Other Clearing Causes are generated by the network, and the standard Diagnostic Codes shown in section 2.17.4 of this appendix are used.

## B 2.17.3 Restart Cause and Diagnostics

Restart Cause field
high order      low order

| | |
|---|---|
| 0 0 0 0 0 0 0 1 | Local Procedure Error |
| 0 0 0 0 0 0 1 1 | Network congestion |
| 0 0 0 0 0 1 1 1 | Network Operational |

Restarting causes are generated by the network, and the standard Diagnostic Codes shown in section 2.17.4 of this appendix are used.

## B 2.17.4 Diagnostic Codes

The following standard Diagnostic Codes are defined. Particular network administrations may use additional codes. The Diagnostic field is optional.

Diagnostic Code byte

| high order | low order | |
|---|---|---|
| 0 0 0 0 | 0 0 0 0 | No additional information to that given in Cause |
| 0 0 0 0 | 0 0 0 1 | P(S) is invalid |
| 0 0 0 0 | 0 0 1 0 | P(R) is invalid |
| 0 0 0 1 | 0 0 0 0 | \| |
| ...to... | | \| This packet type is not valid in the current situation |
| 0 0 0 1 | 1 1 0 1 | \| |
| 0 0 1 0 | 0 0 0 0 | This packet type is not allowed |
| 0 0 1 0 | 0 0 0 1 | Packet cannot be identified |
| 0 0 1 0 | 0 0 1 0 | Call attempt on a one-way LCN |
| 0 0 1 0 | 0 0 1 1 | Invalid packet type on PVC |
| 0 0 1 0 | 0 1 0 0 | Packet on LCN not in use (no Call request) |
| 0 0 1 0 | 0 1 0 1 | Use of REJ packet not agreed with administration. |
| 0 0 1 0 | 0 1 1 0 | Not enough fields for packet |
| 0 0 1 0 | 0 1 1 1 | Too many fields for packet |
| 0 0 1 0 | 1 0 0 0 | Invalid GFI |
| 0 0 1 0 | 1 0 0 1 | RESTART with non-zero LCG or LCN |
| 0 0 1 0 | 1 0 1 0 | Incompatible packet type and facility |
| 0 0 1 0 | 1 0 1 1 | Unauthorized Interrupt Confirmation |
| 0 0 1 0 | 1 1 0 0 | Unauthorized Interrupt |
| 0 0 1 1 | 0 0 0 0 | Timer expiry |
| 0 0 1 1 | 0 0 0 1 | Timer expired on Incoming Call |
| 0 0 1 1 | 0 0 1 0 | Timer expired on Clear Indication |
| 0 0 1 1 | 0 0 1 1 | Timer expired on Reset Indication |
| 0 0 1 1 | 0 1 0 0 | Timer expired on Restart Indication |
| 0 1 0 0 | 0 0 0 0 | Could not establish Call |
| 0 1 0 0 | 0 0 0 1 | Facility code not allowed |
| 0 1 0 0 | 0 0 1 0 | Facility parameter not allowed |
| 0 1 0 0 | 0 0 1 1 | Called address not valid |
| 0 1 0 0 | 0 1 0 0 | Calling address not valid |

# APPENDIX C

## International Alphabet Five (IA5) Character set

| Binary | Hex | Char | Binary | Hex | Char | Binary | Hex | Char | Binary | Hex | Char |
|--------|-----|------|--------|-----|------|--------|-----|------|--------|-----|------|
| 0000 0000 | 00 | NUL | 0010 0000 | 20 | SP | 0100 0000 | 40 | @ | 0110 0000 | 60 | |
| 0000 0001 | 01 | SOH | 0010 0001 | 21 | ! | 0100 0001 | 41 | A | 0110 0001 | 61 | a |
| 0000 0010 | 02 | STX | 0010 0010 | 22 | " | 0100 0010 | 42 | B | 0110 0010 | 62 | b |
| 0000 0011 | 03 | ETX | 0010 0011 | 23 | # | 0100 0011 | 43 | C | 0110 0011 | 63 | c |
| 0000 0100 | 04 | EOT | 0010 0100 | 24 | $ | 0100 0100 | 44 | D | 0110 0100 | 64 | d |
| 0000 0101 | 05 | ENQ | 0010 0101 | 25 | % | 0100 0101 | 45 | E | 0110 0101 | 65 | e |
| 0000 0110 | 06 | ACK | 0010 0110 | 26 | & | 0100 0110 | 46 | F | 0110 0110 | 66 | f |
| 0000 0111 | 07 | BEL | 0010 0111 | 27 | ' | 0100 0111 | 47 | G | 0110 0111 | 67 | g |
| 0000 1000 | 08 | BS | 0010 1000 | 28 | ( | 0100 1000 | 48 | H | 0110 1000 | 68 | h |
| 0000 1001 | 09 | HT | 0010 1001 | 29 | ) | 0100 1001 | 49 | I | 0110 1001 | 69 | i |
| 0000 1010 | 0A | LF | 0010 1010 | 2A | * | 0100 1010 | 4A | J | 0110 1010 | 6A | j |
| 0000 1011 | 0B | VT | 0010 1011 | 2B | + | 0100 1011 | 4B | K | 0110 1011 | 6B | k |
| 0000 1100 | 0C | FF | 0010 1100 | 2C | , | 0100 1100 | 4C | L | 0110 1100 | 6C | l |
| 0000 1101 | 0D | CR | 0010 1101 | 2D | - | 0100 1101 | 4D | M | 0110 1101 | 6D | m |
| 0000 1110 | 0E | SO | 0010 1110 | 2E | . | 0100 1110 | 4E | N | 0110 1110 | 6E | n |
| 0000 1111 | 0F | SI | 0010 1111 | 2F | / | 0100 1111 | 4F | O | 0110 1111 | 6F | o |
| 0001 0000 | 10 | DLE | 0011 0000 | 30 | 0 | 0101 0000 | 50 | P | 0111 0000 | 70 | p |
| 0001 0001 | 11 | XON | 0011 0001 | 31 | 1 | 0101 0001 | 51 | Q | 0111 0001 | 71 | q |
| 0001 0010 | 12 | DC2 | 0011 0010 | 32 | 2 | 0101 0010 | 52 | R | 0111 0010 | 72 | r |
| 0001 0011 | 13 | XOFF | 0011 0011 | 33 | 3 | 0101 0011 | 53 | S | 0111 0011 | 73 | s |
| 0001 0100 | 14 | DC4 | 0011 0100 | 34 | 4 | 0101 0100 | 54 | T | 0111 0100 | 74 | t |
| 0001 0101 | 15 | NAK | 0011 0101 | 35 | 5 | 0101 0101 | 55 | U | 0111 0101 | 75 | u |
| 0001 0110 | 16 | SYN | 0011 0110 | 36 | 6 | 0101 0110 | 56 | V | 0111 0110 | 76 | v |
| 0001 0111 | 17 | ETB | 0011 0111 | 37 | 7 | 0101 0111 | 57 | W | 0111 0111 | 77 | w |
| 0001 1000 | 18 | CAN | 0011 1000 | 38 | 8 | 0101 1000 | 58 | X | 0111 1000 | 78 | x |
| 0001 1001 | 19 | EM | 0011 1001 | 39 | 9 | 0101 1001 | 59 | Y | 0111 1001 | 79 | y |
| 0001 1010 | 1A | SUB | 0011 1010 | 3A | : | 0101 1010 | 5A | Z | 0111 1010 | 7A | z |
| 0001 1011 | 1B | ESC | 0011 1011 | 3B | ; | 0101 1011 | 5B | [ | 0111 1011 | 7B | { |
| 0001 1100 | 1C | FS | 0011 1100 | 3C | < | 0101 1100 | 5C | \ | 0111 1100 | 7C | | |
| 0001 1101 | 1D | GS | 0011 1101 | 3D | = | 0101 1101 | 5D | ] | 0111 1101 | 7D | } |
| 0001 1110 | 1E | RS | 0011 1110 | 3E | > | 0101 1110 | 5E | ^ | 0111 1110 | 7E | ~ |
| 0001 1111 | 1F | US | 0011 1111 | 3F | ? | 0101 1111 | 5F | _ | 0111 1111 | 7F | DEL |

## Meanings of abbreviations

NUL = NULl
SOH = Start Of Heading
STX = Start of TeXt
ETX = End of TeXt
EOT = End Of Trans.
ENQ = ENQuiry
ACK = ACKnowledge
BEL = BELl
BS  = BackSpace
HT  = Horizontal Tab
LF  = Line Feed

VT  = Vertical Tab
FF  = Form Feed
CR  = Carriage Return
SO  = Shift Out
SI  = Shift In
DLE = Data Link Escape
XON = enable data flow
DC2 = Device Control 2
XOFF= stop data flow
DC4 = Device Control 4
NAK = Negative ACK

SYN = SYNchronous idle
ETB = End Trans. Block
CAN = CANcel
EM  = End of Medium
SUB = SUBstitute char.
ESC = ESCape
FS  = File Separator
GS  = Group Separator
RS  = Record Separator
US  = Unit Separator
SP  = SPace
DEL = DELete

**Notes**

IA5 is defined in CCITT recommendation V.3

The high order bit is set according to the parity requirements

XON is sometimes referred to by the more general name of Device Control 1 (DC1), similarly XOFF is sometimes referred to as DC3

The American Standard Code for Information Interchange (ASCII) is usually taken as being the same as IA5. ASCII is not so definite as IA5 since it is very common for manufacturers to replace some of the characters with more useful ones

## APPENDIX D

**Addresses**

Camtec Electronics Ltd.
Camtec House
101 Vaughan Way
LEICESTER
LE1 4SA
England
Tel. 44-533-537534

International Telegraph & Telephone Consultative Committee (CCITT)
General Secretariat
International Telecommunications Union
Place des Nations
CH-1211 Geneva 20
Switzerland
Tel. 41-22-99-51-11

International Organization for Standardization (ISO)
Central Secretariat
1 Rue de Varembe
CH-1211 Geneva 20
Switzerland
Tel. 41-22-34-12-40

Cambridge Ring (CR82) is defined in two documents: the Cambridge Ring 82 Interface Specifications; and the Cambridge Ring 82 Protocol Specifications. These were prepared by the Science and Engineering Research Council, and the Joint Network Team of the Computer Board and Research Councils. Copies can be obtained from the following addresses:

Computing Systems R&D Section
Computing Centre
University of Salford
SALFORD
M5 4WT
England

Joint Network Team
c/o Rutherford Appleton Laboratory
Chilton
DIDCOT
Oxford
OX11 0QX
England

# Index